Analysis of Speedup in Distributed Algorithms

Computer Science: Distributed Database Systems, No. 14

Harold S. Stone, Series Editor
Professor of Electrical and Computer Engineering
University of Massachusetts, Amherst

Other Titles in This Series

No. 7   *Data Management on Distributed Databases*   Benjamin Wan-San Wah

No. 8   *Update Synchronization in Multiaccess Systems*   Milan Milenković

No. 9   *Data Security and Performance Overhead in a Distributed Architecture System*   John M. Cary

No. 10 *Efficiency Analysis of File Organization and Information Retrieval*   J. Dennis Omlor

No. 11 *Search Mechanisms for Large Files*   Marie-Anne Kamal Neimat

No. 12 *Optimization of Queries in Relational Databases*   Yehoshua C. Sagiv

No. 13 *Query Processing Techniques for Distributed Relational Data Base Systems*   Robert S. Epstein

Analysis of Speedup in Distributed Algorithms

by
John P. Fishburn

UMI RESEARCH PRESS
Ann Arbor, Michigan

Copyright © 1984, 1981
John Philip Fishburn
All rights reserved

Produced and distributed by
UMI Research Press
an imprint of
University Microfilms International
Ann Arbor, Michigan 48106

Library of Congress Cataloging in Publication Data

Fishburn, John P.
Analysis of speedup in distributed algorithms.

   (Computer science. Distributed database systems ; no. 14)
   Revision of thesis (Ph.D)–University of Wisconsin–Madison, 1981.
     Bibliography: p.
     Includes index.
     1. Electronic data processing–Distributed processing.
2. Parallel processing (Electronic computers)
3. Algorithms. I. Title. II. Series.

QA76.9.D5F57 1984   001.64   83-18307
ISBN 0-8357-1527-2

Contents

Acknowledgments ix

1 Introduction 1

2 Judging Parallel Algorithms 3

3 Previous Research 5
    3.1. Sorting and Merging 5
        3.1.1. Comparison-Exchange Networks 5
        3.1.2. Parallel Tape Sorting 6
        3.1.3. Multi-Processor Methods 6
        3.1.4. SIMD Methods 6
        3.1.5. Vector Sorting 7
    3.2. Numerical Methods 7
        3.2.1. Fast-Fourier Transform 7
        3.2.2 Adaptive Quadrature 7
        3.2.3. Matrix Methods 7
    3.3. Global Structuring 8
    3.4. Graph Theory 9

4 Parallel Alpha-Beta Search 11
    4.1 Introduction 11
    4.2. The Alpha-Beta Algorithm 12
    4.3. Related Work 15
        4.3.1. Parallel-Aspiration Search 15
        4.3.2. Mandatory-Work-First Search 15
    4.4. The Tree-Splitting Algorithm 16
        4.4.1. The Leaf Algorithm 16
        4.4.2. The Interior Algorithm 17
        4.4.3. Alpha Raising 19
    4.5. Measurements of the Algorithm 19

4.6. Optimizations 21
4.7. Analysis of Speedup 22
    4.7.1. Worst-First Ordering 23
    4.7.2. Best-First Ordering 24
    4.7.3. Discussion 28
    4.7.4. Random Order 29
    4.7.5. Discussion of Theorem 4.6 39
4.8. Mandatory-Work-First Search 41
    4.8.1. Best-First Order 44
    4.8.2. Worst-First Order 47
    4.8.3. Other Orderings 49
4.9. Comparison of Palphabeta and mwf 50
4.10. Tips for Processor-Tree Architects 51
    4.10.1. Serial versus Parallel 51
    4.10.2. Maximal Processor Trees 53

5 Piecewise-Serial Iterative Methods 55
5.1. Introduction 55
5.2. The Dirichlet Problem 55
    5.2.1. Jacobi Method 56
    5.2.2. Gauss-Seidel Method 57
    5.2.3. Successive Over-Relaxation 57
5.3. Previous Work 58
5.4. Piecewise-Serial Iterative Methods 59
    5.4.1. Uniform Regions With Grid Topology 59
    5.4.2. Uniform Regions With Tree Topology 61
    5.4.3. Efficiency 71
    5.4.4. Measurement of Communication Time 72
    5.4.5. Scheduling Tree Machines 74
    5.4.6. Static Nonuniform Regions 75
    5.4.7. Dynamic Region Encroachment 75

6 Quotient Networks 83
6.1. Introduction 83
6.2. Existing Networks 83
    6.2.1. Grid-Connected Network 84
    6.2.2. Perfect Shuffle 84
    6.2.3. PM2I 84
    6.2.4. Cube 85
6.3. Existing Algorithms 85
    6.3.1. Fast-Fourier Transform on the Shuffle 85
    6.3.2. Sorting on the Shuffle 87

      6.3.3. Polynomial Evaluation on the Shuffle   88
      6.3.4. Finite-Difference Methods   88
  6.4. Network Emulation   88
      6.4.1. Perfect Shuffle   89
      6.4.2. Grid-Connected Network   90
      6.4.3. Cube   93
      6.4.4. PM2I   94
  6.5. Some Resulting Algorithms   97
      6.5.1. Fast-Fourier Transform on the Shuffle   97
      6.5.2. Sorting on the Shuffle   98
      6.5.3. Polynomial Evaluation on the Shuffle   99
      6.5.4. Finite-Difference Methods   99
      6.5.5. Alpha-Beta Search   99
  6.6. The Economics of Emulation   99

7 Conclusions and Future Directions   101
  7.1. Alpha-Beta Search   101
  7.2. Piecewise-Serial Iterative Methods   102
  7.3. Quotient Networks   103
  7.4. Parallel Programming Proverbs   104
      7.4.1. Large Computation per Message   104
      7.4.2. Do Interesting Work First   104
      7.4.3. Do Mandatory Work First   105
      7.4.4. Do Something   105

Appendix A: Some Optimizations of α–β Search   107

Bibliography   113

Index   117

Acknowledgments

Thanks go first to Raphael Finkel, who spent an enormous amount of energy editing rough drafts and prodding me in fruitful directions.

Jim Goodman, Will Leland, Diane Smith, Marvin Solomon, John Strikwerda, and Len Uhr have provided many helpful discussions. Karl Anderson started me on parallel alpha-beta search by directing me to Baudet's work. I am grateful to these people and thank them all.

Finally, I want to give special thanks to Sharon Lawless for writing part of the Arachne checkers program, Mary Leland for providing Lemma 4.5, Karen Desiato for careful proofreading and Raphael Finkel for translating this research into English.

1

Introduction

Three helping one another will do as much as six men singly.

Spanish Proverb

SHUTTLE DELAYED; COMPUTERS WOULDN'T TALK

The Capital Times
10 April 1981

Most computers consist of a single central processing unit (CPU), a store of words, and a communications line between them. A program's task is to change the store's content in some significant way. It accomplishes this change by passing information along the line, one word at a time, back and forth between the CPU and store. Because of its serial nature, Backus [1] calls this line the "von Neumann bottleneck".

The von Neumann bottleneck not only limits the speed of ordinary computers, but also forces us to think of algorithms in serial terms. In recent years, several computer architectures have been proposed that avoid the von Neumann bottleneck by allowing many computations to proceed simultaneously. Some of these architectures have been built and are working [2,3,4,5,6,7].

We use the taxonomy of Flynn [8] to divide parallel processors into two broad classes: *MIMD* and *SIMD*. In the MIMD (Multiple Instruction stream, Multiple Data stream) model, each processor is a separate computer with its own program counter. Each processor computes independently of all others. MIMD computers can be broken down into subclasses: A *multicomputer* consists of several ordinary computers connected only by communications lines. Arachne [7] is a multicomputer. A *multiprocessor* consists of several CPUs with shared access to a common memory. C.mmp [3] is a multiprocessor.

In an SIMD network, a central controller broadcasts one instruction at a time to all the processors in the network, which then execute the instruction simultaneously on their own data. Illiac IV [2] is an SIMD network.

2 Introduction

If parallel architectures are ever to become widely useful, we must learn how to use them efficiently. In this study we investigate the use of parallel architectures in performing certain computations. We assume throughout that each processing unit has a private memory and is connected by communications lines to some of the other processors. Communication is restricted to data passed on these lines; no shared memory exists in our model. Although we usually assume the multicomputer model, chapters 5 and 6 deal with interconnection networks that may be SIMD or MIMD.

Chapter 2 discusses figures of merit for parallel algorithms. Chapter 3 briefly surveys previous work in the field of parallel algorithms. Chapter 4 presents two parallel alpha-beta search algorithms. The alpha-beta pruning technique is used by programs that play games such as chess to speed up the search of the tree of possible continuations. Alpha-beta search presents a challenge to the designer of parallel algorithms because of its inherently serial nature: Results from searching one part of the lookahead tree reduce the computation for searching another part. If both searches proceed independently, these savings are reduced.

Chapter 5 presents several parallel implementations of the Jacobi method. The Jacobi method is an important technique for numerically solving certain partial differential equations such as the Dirichlet problem.

A large-network algorithm solves a problem of size N on an interconnection network of N processors. A general method for transforming large-network algorithms into quotient-network algorithms, which solve problems of size N on networks with fewer processors, is presented in chapter 6. This transformation allows algorithms to be designed for certain interconnection topologies assuming any number of processing elements.

Chapter 7 summarizes the contributions of this study and discusses areas for further work.

2

Judging Parallel Algorithms

When the judge is unjust he is no longer a judge but a transgressor.

Phillips Brooks
Visions and Tasks

By what standards can we judge parallel algorithms? The most commonly used gauge of a parallel algorithm's performance is *speedup*, which we define as:

$$\text{speedup} = \frac{\text{time required by the best serial algorithm}}{\text{time required by the parallel algorithm}}$$

The running time for the parallel algorithm includes time required for data movement. Sometimes speedup is of overwhelming importance. For example, if a 24-hour weather-prediction program that runs serially in 48 hours could be made to run four times faster with a tenfold increase in hardware, such a conversion might very well be considered appropriate. Whenever someone must wait for a program to complete, we may be willing to pay for more than an N-fold increase in hardware to obtain an N-fold speedup. Examples of such computations are database transactions and work performed for interactive users.

Another criterion used to judge parallel algorithms is *efficiency*:

$$\text{efficiency} = \frac{\text{speedup}}{\text{number of processors used}}$$

The use of this criterion assumes that cost is proportional to the number of processors. This assumption is sometimes overoptimistic, as for example in architectures that use a crosspoint switch of complexity N^2 to connect N processors to N memories.

Sometimes the efficiency of a parallel algorithm is greater than one, which leads us to conclude that the serial algorithm to which it is compared is not the best available. For example, Baudet [9] found, for k equal to two or three, more than k-fold speedup in performing alpha-beta search with k processors. However, the serial algorithm under comparison unwisely started the search with the window (-infinity,+infinity), instead of the usual narrow window. (This algorithm is described in detail in chap. 4.)

The designer of parallel algorithms hopes to achieve an efficiency of one. Some algorithms achieve this (e.g., Pease's use of the perfect shuffle [10] to compute FFTs), others come close (the efficiency of Batcher's sorting algorithm [11] is $1/\log N$), others fall short (Csanky's algorithm [12] computes the inverse of a matrix with efficiency $1/N\log^2 N$).

A third criterion is *practicality*, by which we mean the likelihood that the required hardware will exist at some time in the near future. Many of the algorithms reviewed in chapter 3 are quite impractical. First, many of these algorithms assume that all processors have equal and noninterfering access to a common memory. Workable hardware that fits this description is both expensive and inefficient, especially when the number of processors is large. (The best example is C.mmp [3], which uses an expensive crossbar switch to connect processors to memory, and has serious problems with memory contention.) Common memory places a practical upper bound on the number of processors that can be used. In effect, the von Neumann bottleneck on a shared memory machine must serve many processors instead of one. Second, many parallel algorithms require N (or worse, N^2) processors to solve a problem of size N.

3

Previous Research

The average Ph.D. thesis is nothing but a transference of bones from one graveyard to another.

J. Frank Dobie

The body of literature on parallel algorithms is growing rapidly. Most of it assumes a C.mmp-like (MIMD, shared-memory) architecture. In this section we review previous results in parallel algorithms for sorting, numerical methods, global structuring, and graph theory.

3.1. Sorting and Merging

Previous work in parallel sorting methods can be divided into five broad categories: comparison-exchange networks, parallel tape sorting, multiprocessor methods, SIMD methods, and vector sorting.

3.1.1. Comparison-Exchange Networks

A comparison-exchange network accepts N numbers on N input lines and sorts them onto N output lines by means of a network of *comparison-exchange modules*. After receiving two numbers on its two input lines, a comparison-exchange module always places the larger number on a particular output line and the smaller on the other output line. A sorting network, as we shall call it, can also be thought of as a multiprocessor architecture with the following nonadaptivity constraint: Whenever two numbers K_i and K_j are compared, the subsequent comparisons in the case $K_i < K_j$ are identical to those in the case $K_i > K_j$, except that i and j are interchanged. Batcher, in one of the earliest results on sorting networks [11], shows how to sort N words in $(1/2)\log N(\log N + 1)$ steps with approximately $(1/2)\log^2 N$ ranks of $N/2$ modules each. Stone [13], building on Batcher's work, accomplishes the same task with only one rank of modules. Muller [14], by allowing the network to contain AND and OR gates and single-pole, double-

6 Previous Research

throw switches, as well as the comparator modules, shows how to sort N words in time O(log N). Unfortunately, this scheme uses $O(N^2)$ elements. Knuth [15] gives an excellent, but dated, review of sorting networks.

3.1.2. Parallel Tape Sorting

Parallel tape sorting is a parallel version of external tape sorting. Parallel tape sorting exploits the fact that, after information has been placed on a tape by one computer, the tape can be used immediately as input to another computer. This technique is unique among parallel sorting methods in not assuming special hardware. Given N records, log N processors, and 4(log N) tapes, Even [16] gives a method that sorts the records onto a tape in time 3N−2, where it is assumed that a record can be read from tape and written onto another tape in one time unit.

3.1.3. Multi-Processor Methods

Multi-processor methods use the architectural model of many processors with equal access to a common memory. As with most algorithms, most of the work done in parallel sorting schemes assumes this architecture. Valiant [17] gives an algorithm for sorting N words with N processors in 2logN · loglogN + O(log N) comparison steps. Gavril [18] gives an algorithm that merges two linearly ordered sets of size N, with $P \leq N$ processors in 2[log(N + 1)] + (4N/P) steps. Hirschberg [19] gives an algorithm that sorts N words in time O(Klog N) using $N^{1 + 1/k}$ processors, for k an arbitrary integer. Unfortunately, the amount of memory necessary is MN, where the numbers to be sorted are in the range [0,M−1]. Furthermore, memory fetch conflicts do exist. That is, the algorithm assumes an architecture that, in one time unit, can satisfy multiple read requests to a single memory cell. Preparata [20] improves on Hirschberg's algorithms by giving a family with identical performance, but without memory fetch conflicts.

3.1.4. SIMD Methods

Baudet [21] gives an algorithm for sorting N words on K processors in time (Nlog N)/K + O(N). Hence, when K ≪ log N, the speedup is optimal in the number of processors used. Baudet's method can thus be considered practical, since performance can be boosted linearly with a small number of processors. Thompson [22], on the other hand, gives two algorithms for sorting N^2 words on an N-by-N mesh-connected processor array (like the Illiac IV) in O(N) routing and comparison steps.

3.1.5. Vector Sorting

Stone [23] studies several different sorting methods for one particular architecture, the CDC STAR computer. He shows that although the $N(\log^2 N)$ computational complexity of Batcher's bitonic sorting algorithm is worse than Quicksort's, the bitonic sort's good use of STAR's vector instructions allows it to out-perform Quicksort on vectors of reasonable size.

3.2. Numerical Methods

This section reviews parallel numerical algorithms that have been developed for the fast-Fourier transform, zero-finding, adaptive quadrature, and matrix computations. A review of parallel algorithms for finite-difference calculations is given in chapter 5.

3.2.1. Fast-Fourier Transform

One of the most important discoveries in algorithms in recent years has been the fast-Fourier transform (FFT) [24]. Pease [10] demonstrates that the perfect shuffle interconnection pattern can yield optimal speedup in the computation of the DFT. Specifically, he shows that log N passes through N/2 *multiply-add* modules, alternating with log N passes through an N-line shuffle-exchange, is sufficient to compute the DFT of N points in time $O(\log N)$. (Both the FFT and the shuffle-exchange network are discussed in detail in chap. 6.)

Flanders [6] shows how to accomplish the necessary routing for the computation of a DFT on a rectangular grid of processors.

3.2.2. Adaptive Quadrature

Lemme [25] describes a parallel architecture for calculating finite-sum estimates of one-dimensional integrals. The architecture consists of a tree of computers connected by communications lines. In addition, all leaf processors have access to a large common memory that specifies a queue of tasks for each processor. These queues are managed by a set of queue-balancing processors. The speedup associated with this configuration is shown to be at least order of $N/\log N$ with N processors.

3.2.3. Matrix Methods

Parallel algorithms have been developed for solving tridiagonal systems, band triangular systems, and for matrix inversion and matrix multiplication.

8 Previous Research

Traub [26] and Stone [27] both consider the solution of equations whose matrix has nonzero elements only on the three central diagonals. Traub proposes an iterative method, called Parallel Gauss, and shows that m processors can solve a linear system of size m in time $O(1)$. The Parallel Gauss method can be run on SIMD machines (such as the Illiac IV), or on C.mmp-like machines. Stone [28] presents an iterative method, called the odd-even reduction algorithm, that under diagonal-dominance conditions converges more quickly than Traub's Parallel Gauss method.

Chen [29] considers the solution of lower-triangular systems of equations. Chen gives direct methods for solving these systems, and shows that when the bandwidth of the matrix is m+1 (diagonals further than m away from the main diagonal are all zero), these methods yield a speedup of approximately p/m with p processors.

Gentleman [30], by considering only data movement, gives lower bounds to the parallel complexity of certain matrix operations. In particular, he shows that for machines with two-dimensional rectangular grid connectivity (like the Illiac IV), multiplication and inversion of N-by-N matrices inherently require order of N steps.

Csanky [12] gives an algorithm that computes the inverse of an N-by-N matrix in time $O(\log^2 N)$ using $O(N^4)$ processors. Preparata [31], by modifying Csanky's algorithm, shows that the same time bound can be achieved with $2N^{A+(1/2)}/(\log^2 N)$ processors if multiplication of two N-by-N matrices can be done in parallel in time $O(\log N)$ using $N^A/\log N$ processors.

3.3. Global Structuring

Finkel [32] empirically investigates distributed algorithms for imposing global structure on graphs. Pairing algorithms match together in pairs as many neighboring nodes as possible. Spanning tree algorithms select a subset of the edges that provides a unique path between any two nodes. Finally, developing hierarchies generalizes the pairing algorithms in two ways: First, nodes are organized into groups of arbitrary size, rather than in pairs. Second, these groups are then treated as nodes, to be grouped into meta-groups, and so on. These algorithms assume the graph to be identical to the physical network topology. Moreover, information about the structure of the graph is itself distributed: Each processor knows only of its own immediate neighbors. These algorithms are therefore suitable not for processing arbitrary graphs on arbitrary networks, but for organizing a given physical network as a basis for solving other problems.

Change [33] assumes similar ground rules, and addresses the tasks of finding a minimal spanning tree of a weighted graph, distributing a list, and finding the extrema of a set of nodes, given that their names obey a total ordering.

3.4. Graph Theory

Most work in parallel graph-theoretic algorithms assumes an architectural model of many computers with equal access to a large common memory. Given an adjacency matrix as input, Hirschberg [34] gives a parallel algorithm that uses N^2 processors to compute the connected components of an undirected graph with N nodes in time $O(\log^2 N)$. Savage [35] presents a family of algorithms that use $O(\log^2 N)$ running time, and polynomial-in-N processors, to solve the following problems for a graph with N nodes: For a connected, undirected graph G, find a spanning tree, a cycle, a cycle basis, the bridges and bridge connected components, and the biconnected components of G. For a connected, undirected, weighted graph, find a minimum spanning tree. For a connected, directed graph, find a cycle, a shortest cycle, the dominators, and the dominator tree.

4

Parallel Alpha-Beta Search

The axe is already laid at the root of the trees; so every tree that fails to yield good fruit will be cut down and thrown into the fire.

John the Baptist

> *By the Nine Gods he swore it,*
> *And named a trysting day,*
> *And bade his messengers ride forth*
> *East and west and south and north,*
> *To summon his array.*

"Lays of Ancient Rome"
Lord Macaulay

4.1. Introduction

The α–β search algorithm is central to most programs that play games like chess. It is now well known [36] that an important component of the playing skill of such programs is the speed at which the search is conducted. For a given amount of computing time, a faster search allows the program to "see" farther into the future. This chapter presents and analyzes two parallel adaptations of the α–β algorithm. The first adaptation, called the *tree-splitting algorithm*, speeds up the search of a large tree of potential continuations by dynamically assigning subtree searches for parallel execution.

Section 2 reviews the α–β algorithm. Section 3 discusses parallel implementations of the α–β algorithm suggested by other workers. Section 4 formally describes the tree-splitting algorithm. Section 5 presents performance measurements for this algorithm taken on a network of microprocessors. Section 6 discusses some possible optimizations and variations of the algorithm. Section 7 derives the obtainable speedup with k processors as k tends towards ∞.

The second adaptation, *mandatory work first*, is a formalization of a method proposed by Akl, Barnard, and Doran [37]. Section 8 analyzes this algorithm. Section 9 compares it with the tree-splitting algorithm. A number of suggestions for architectural design of processor trees closes this chapter.

4.2. The Alpha-Beta Algorithm

Consider a board position from a game such as chess or checkers. All possible sequences of moves from this position may be represented by a tree of positions called the *lookahead tree* (fig. 4.1). The nodes of the tree represent positions; the children of a node are moves from that node. The root node of the tree represents the current position. Since lookahead trees for most games are often too large to be searched even by computer, they are usually truncated at a certain level. Since we will later be referring to a tree of processors, we reserve the following notation for nodes of lookahead trees: A node is often called a *position*. A node's child is its *successor*, and its parent is its *predecessor*. If each interior node has n successors, we say that the tree has *degree n*. The *level* of a node or subtree is its distance from the root.

The α-β algorithm is an optimization of the minimax algorithm, which we will review first. The two players are called *max* and *min*; at the root node, it is max's turn to move. The minimax algorithm proceeds as follows: First, each leaf of the lookahead tree is assigned a *static value* that reflects that position's desirability. (High values are desirable to max. In a game such as chess, the main component of the value is usually the material balance between the two sides.)

The interior nodes of the lookahead tree may be given *minimax values* recursively: If it is max's turn to move at node A, the value of A is the maximum of A's successors' values. If the game were to proceed to node A, it

Fig. 4.1. Lookahead tree.

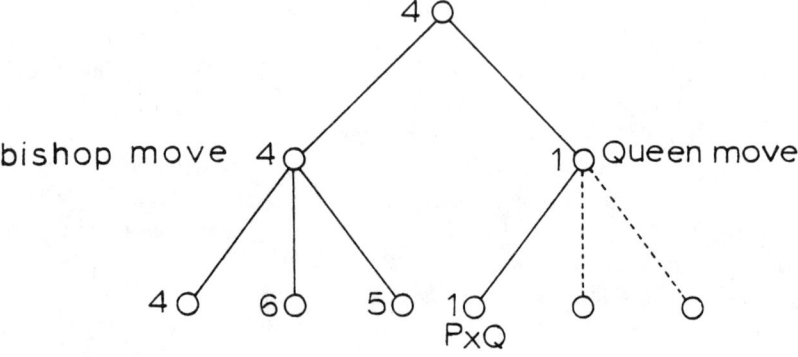

would then be max's turn to move. Max, being rational, would choose the successor with the maximum value, say M. Therefore, the subtree rooted at A must have M as its value, because M is the value of the leaf node we would reach if the game reached A. Similarly, if it is min's turn to move at a node, then the value of that node is the minimum of these values.

We will use a version of the minimax procedure called *negamax*: When it is max's turn to move at a terminal node, the node is assigned the same static value used in minimax. When it is min's turn to move, the static value assigned is the negative of what it would be in the minimax case. The value of an interior node at any level is defined to be the maximum of the negatives of the values of its successors.

The negamax algorithm can be cast into an *ad hoc* Pascal-like language. The following program is adapted from Knuth [38]:

```
function negamax(p:position):integer;
var  m:  integer;
     i,d : 1..MAXCHILD;
     succ : array[1..MAXCHILD] of position;
begin
   determine the successor positions
      succ[1], ..., succ[d];
   if d = 0 then { terminal node }
      negamax := staticvalue(p)
   else
   begin { find maximum of child values }
      m := - ∞;
      for i := 1 to d do
         m := max(m,- negamax(succ[i]));
      return(m);
   end
end.
```

The α–β algorithm evaluates the lookahead tree without pursuing irrelevant branches. Suppose we are investigating the successors in a game of chess, and the first move we look at is a bishop move. After analyzing it, we decide that it will gain us a pawn. Next we consider a queen move. In considering our opponent's replies to the queen move, we discover one that can irrefutably capture the queen; she has moved to a dangerous spot. We need not investigate our opponent's remaining replies; in light of the worth of the bishop move, the queen move is already discredited.

The α–β search algorithm [38] formalizes this notion:

```
function alphabeta(p : position; α,β : integer) : integer;
var i,d : 1..MAXCHILD;
    succ : array [1..MAXCHILD] of position;
begin
    determine the successor positions
        succ[1], ..., succ[d];
    if d = 0 then
        alphabeta := staticvalue(p)
    else
    begin
        for i := 1 to d do
        begin
            α := max(α, - alphabeta(succ[i], -β,-α));
            if α ≥ β then return(α) { cutoff }
        end;
        return(α);
    end
end.
```

The function alphabeta obeys the *accuracy property*: For a given position p, and for values of α and β such that $\alpha < \beta$,

if negamax(p) $\leq \alpha$, then alphabeta(p,α,β) $\leq \alpha$
if negamax(p) $\geq \beta$, then alphabeta(p,α,β) $\geq \beta$
if $\alpha <$ negamax(p) $< \beta$, then alphabeta(p,α,β) = negamax(p)

The first and second cases above are called *failing low* and *failing high* respectively. In the third case, *success*, alphabeta accurately reports the negamax value of the tree. Success is assured if $\alpha = -\infty$ and $\beta = \infty$. The pair (α,β) is called the *window* for the search.

To return to our example: When alphabeta is called with p representing the queen move, it is min's move. β is the cutoff value generated by the bishop move. The better the bishop move was for max, the lower is β. (Within the routine alphabeta, high values for α and β are good for the player whose move it is. A high value for α indicates that a good alternative for that player exists somewhere in the tree. A low value for β indicates that a good alternative exists for the other player somewhere else in the tree.) When the successor that captures the queen is evaluated, α becomes larger than β, and a cutoff occurs.

α-β pruning serves to reduce the *branching factor*, which is the ratio between the number of nodes searched in a tree of height N and one of height N-1, as N tends to ∞. Both theory [38] and practice [39] agree that with good move ordering (investigating best moves first), α-β pruning reduces the branching factor from the degree of the lookahead tree nearly to the square root of that degree. For a given amount of computing time, this reduction nearly doubles the depth of the accessible lookahead tree.

When the algorithm is performed on a serial computer, the value of one successor can be used to save work in evaluating its siblings later on. Nevertheless, greater speed can be obtained by conducting α-β search in a parallel fashion.

We will restrict our attention to parallel computers built as a tree of serial computers. A node in this tree is a *processor*, the parent of a node is its *master*, and the child of a node is its *slave*.

4.3. Related Work

In this section we review previous research in parallel alpha-beta algorithms.

4.3.1. Parallel-Aspiration Search

In order to introduce parallelism, Baudet [9] rejects decomposition of the lookahead tree in favor of a *parallel aspiration search*, in which all slave processors search the entire lookahead tree, but with different initial α-β windows. These windows are disjoint, and in the simplest variant their union covers the range from $-\infty$ to $+\infty$. Since each window is considerably smaller than $(-\infty, +\infty)$, each processor can conduct its search more quickly. When the processor whose window contains the true minimax value of the tree finishes, it reports this value, and move selection is complete. Baudet analyzes several variants of this algorithm under the assumption of randomly distributed terminal values, and concludes that the obtainable speedup is limited by a constant independent of the number of processors available. This maximum is established to be approximately 5 or 6. Surprisingly, for k equal to 2 or 3, Baudet's method yields more than k-way speedup with k processors. Baudet infers that the serial α-β search algorithm is not optimal, and estimates that a 15 to 25 percent speedup may be gained by starting the search with a narrow window.

Since a narrow window does not speed up a successful search when moves are ordered best-first, Baudet's method yields no speedup under best-first move ordering.

4.3.2. Mandatory-Work-First Search

Akl, Barnard, and Doran [37] distinguish between those parts of a subtree that must be searched and those parts whose need to be searched is contingent upon search results in other parts of the tree. By searching mandatory nodes first, their algorithm attempts to achieve as many of the cutoffs seen in the serial case as possible. This technique leads to an algorithm that we discuss in detail in section 4.8.

4.4. The Tree-Splitting Algorithm

A natural way to implement the α–β algorithm on parallel processors divides the lookahead tree into its subtrees at the top level and queues them for parallel assignment to a pool of slave processors. Each processor computes the value of its assigned subtree by using either serial α–β search (if it is a leaf processor) or parallel α–β search (if it has slaves of its own). When it finishes, it reports the value computed to its master. As a master receives responses from its slaves, it narrows its window and tells working slaves about the improved window. When all subtrees have been evaluated, the master is able to compute the value of its position.

4.4.1 The Leaf Algorithm

The leaf algorithm runs at leaf nodes of the processor tree. We will describe its interactions with its master by means of remote procedure calls. The algorithm can also be expressed in a message-passing or shared-memory form. The master calls the function leaf$\alpha\beta$ (line 19) remotely. A master can interrupt a search in progress to tell its slave of a newly narrowed window by invoking the asynchronous "update" procedure in the slave (line 3). The variables α and β (line 1) are global arrays, not formal parameters, in order to facilitate updating their values in each recursive call of alphabeta when the new window arrives. Here is the leaf algorithm:

```
1    α,β : array[1..MAXDEPTH] of integer;

3    asynchronous procedure update(newα, newβ : integer);
4      { update is called asynchronously by my master
5        to inform me of the new window (newα,newβ) }
6    var tmp : integer;
7        k : 1..MAXDEPTH;
8    begin
9      for k := 1 to MAXDEPTH do
10       begin { update α,β arrays }
11         α[k] := max (α[k],newα);
12         β[k] := min (β[k],newβ);
13         tmp := newα;
```

```
14        newα := -newβ;
15        newβ := -tmp;
16      end
17    end;

19    function leafαβ(p : position; α,β : integer) : integer;
20    begin
21      α[1] := α;
22      β[1] := β;
23      return(alphabeta(p,1));
24    end;

26    function alphabeta(p:position; depth:integer): integer;
27    var succ:array [1..MAXCHILD] of position; {successors}
28        succno : 1..MAXCHILD; { which successor }
29        succlim : 1..MAXCHILD; { how many successors }
30    begin
31      determine the successors succ[1], ..., succ[succlim];
32      if succlim = 0 then return(staticvalue(p));
33      for succno := 1 to succlim do
34      begin { evaluate each successor }
35        α[depth+1] := - β[depth];
36        β[depth+1] := - α[depth];
37        α[depth] := max(α[depth],
38          -alphabeta(succ[succno]depth+1));
39        if α[depth] ≥ β[depth] then
40          return(α[depth]); { cutoff occurs }
41      end { for succno }
42      return(α[depth]);
43    end; { function alphabeta }
```

4.4.2 The Interior Algorithm

The interior algorithm interiorαβ runs on interior nodes of the processor tree. When interiorαβ is activated, it generates all successors of the position to be evaluated (line 25). Each of its slaves is requested to evaluate one of these positions; the remaining positions are queued for later service. Newly narrowed windows are relayed to slaves by use of "update" calls (line 3).

The master may take various actions when its slave returns. First, if the returned value causes the current α value to increase, then the master sends $-α$ as an updated β value to all of its active slaves (line 39). Second, if α has been increased so that it becomes greater than or equal to β, then an α-β cutoff occurs. The nonpositive-width window is sent to all active slaves, quickly terminating them (line 39). Meanwhile, the master empties its queue of waiting successor positions. (In the algorithm shown below, this effect is achieved by invoking slaves with negative-width windows.) Third, if the queue of unevaluated successor positions is nonempty, the reporting slave is assigned the next position from the queue.

When all successors have been evaluated, the master returns the final value to its master. In a game situation, the algorithm at the root node might serve as the user interface, and would remember which move has the maximum value.

Here is the interior algorithm:

```
1   var glα,glβ : integer; { global α,β }
2          q : integer; { depth of processor tree }
3   asynchronous procedure update(newα, newβ : integer);
4     { update is called asynchronously by my master
5     to inform me of the new window (newα,newβ) }
6   begin
7     atomically do
8     begin
9        glα := max(glα,newα);
10       glβ := min(glβ,newβ);
11    end; { atomically do }
12    for all slaveid do
13       slaveid.update(-glβ,-glα);
14  end; { update }

16  function interiorαβ(p: position ; α,β: integer) : integer;
17  var succ: array[1..MAXCHILD] of position; { successors }
18      succno : 1..MAXCHILD; { which successor }
19      succlim : 1..MAXCHILD; { how many successors }
20      tmp : array[1..MAXCHILD] of integer;
21      function g : integer;
22  begin
23     glα := α;
24     glβ := β;
25     determine the successors succ[1], ..., succ[succlim];
26     if succlim = 0 then return(staticvalue(p));
27     if depth(succ[1]) < q then
28        g := interiorαβ;
29     else g := leafαβ;
30     parfor succno := 1 to succlim do
31     begin
```

```
32      when slaveid := idleslave() do
33          tmp[succno] :=
34              -slaveid.g(succ[succno],-glβ,-glα);
35          if tmp[succno] > glα then
36          begin
37              atomically do glα := max(tmp[succno],glα);
38              for all slaveid do
39                  slaveid.update(-glβ,-glα);
40          end; { if tmp[succno] > glα }
41      end; { par for succno }
42      return (gld);
43  end; { interior }
```

4.4.3. Alpha Raising

As an optimization of the interior algorithm, the master running on the root node may send a special α-β window to a slave working on the last unevaluated successor. This window is $(-\alpha-1,-\alpha)$ instead of the usual $(-\beta,-\alpha)$. If that successor is not the best, then the slave's search will fail high as usual, but the minimal window speeds its search. If that successor is best, then the smaller window causes the search to fail low, again terminating faster. In either case, the root master determines which successor is the best move, even though its value may not be calculated. By speeding the search of the last successor, the idle time of the other slaves is reduced. (This narrow window given to the root's last subtree search can also be used in serial α-β search, as discussed in the appendix.)

We can generalize this technique in the following way, called *alpha raising*: Suppose that each successor of the root is being evaluated by a different slave, and that slave$_1$'s current α value, α_1, is lower than any other, and that slave$_2$ has the second lowest α value, say α_2. Update α_1 to α_2-1, speeding up slave$_1$. If this update causes slave$_1$'s otherwise successful search to fail low, then the reported value is still lower than all others, and that move is still discovered to be best.

4.5. Measurements of the Algorithm

Measurements of the performance of the tree-splitting algorithm have been taken on a network of LSI-11 microcomputers running under the Arachne [7] operating system.

The game of checkers was used to generate lookahead trees. Static evaluation was based on the difference in a combination of material, central board position for kings and advancement for men. Moves were ordered best-first according to their static values. General α-raising was not employed except for the special case for the last successor.

A single LSI-11 machine searches lookahead trees at a rate of about 100 unpruned nodes per second. Inter-machine messages can be sent at a rate of about 70 per second.

Only five processors were available in Arachne at the time of these experiments, so it was not possible to directly test processor trees of height greater than one. An estimate of the speedup of a tree of height two was made by exploiting the following fact: Since a master spends most of its time waiting for its slaves to finish their assigned tasks, the speed of a master is proportional to the speed of its slaves. One way to speed up a leaf processor is to replace it with a processor tree of height one. Therefore we can roughly equate the speedup of a height-two processor tree in searching a height-x lookahead tree with the product $\gamma_0\gamma_1$, where γ_0 is the speedup of a height-one processor tree in searching a height-x lookahead tree, and γ_1 is the speedup of a height-one processor tree in searching a lookahead tree of height x-1.

Ten board positions, $B_1, \ldots, B_{10}$, were chosen for use in these experiments. These positions actually arose during a human-machine game; they span the entire game. All lookahead trees from these positions were expanded to a depth of eight.

Two sets of experiments were performed. The two differed only in that the first set used one master and two slaves, while the second set used one master and three slaves. Within each experiment, γ_0 was measured directly for each B_i by evaluating the tree both serially and with the parallel algorithm running on a depth-one processor tree. Table 1 summarizes measurements of γ_0.

The 10 board positions gave rise to 84 successors, so 84 EVALUATE commands were given to slaves while γ_0 was being measured. These 84 commands were saved, and times for both parallel and serial evaluation were measured for each command. The *aggregate speedup* for a group of commands is the total time required to execute them serially divided by the total time required to execute them in parallel. For each B_i, the aggregate speedup γ_1 for its subtree evaluations was computed. Table 2 summarizes measurements of γ_1.

Table 1. γ_0 for each B_i, i=1,..., 10

	2 slaves	3 slaves
minimum	1.37	1.37
average	1.81	2.34
maximum	2.36	3.15
standard deviation	0.31	0.56

Table 2. γ_1 for each B_i, i=1,...,10

	2 slaves	3 slaves
minimum	1.03	1.38
average	1.46	1.96
maximum	1.77	2.60
standard deviation	0.22	0.38

Surprisingly, more than k-way speedup occasionally was achieved with k slaves: Three out of the 10 B_i were sped up by more than 2 with 2 slaves, and two of those three B_i were sped up by more than 3 with 3 slaves. Of the 84 subtrees of the B_is, 4 were sped up by more than 2 with 2 slaves, and 9 were sped up by more than 3 with 3 slaves; 2 of those achieved 6-way speedup. In each such case, subtree evaluations finished in a different order than they were assigned. While one large subtree was being evaluated by one slave, another smaller subtree was assigned and finished. The large subtree's evaluation then received an UPDATE message that sped it up or even terminated it. In fact, time-consuming searches are more likely than short ones to receive these messages. In particular, the search that receives the final $(-\alpha-1,-\alpha)$ window is likely to be larger than average.

4.6. Optimizations

Since the tree-splitting algorithm can be optimized in several ways, it should be considered the simplest variant of a family of tree-decomposing algorithms for $\alpha-\beta$ search. As a first optimization, since most of a master's time is spent waiting for messages, that time could be spent profitably doing subtree searches. However, only the deepest masters could hope to compete with their slaves in conducting searches. All other masters are by themselves slower than their slaves because their slaves have slaves below them to help. However, more than half of all masters control leaf processors, and greater speedup should be achieved by running a leaf algorithm along with these masters on the same processors. We might expect an additional 1.5-way speedup from this technique.

A second optimization groups several higher-level masters onto a single processor. For example, the three highest processors in a binary processor tree could be replaced by three processes running on a single processor.

Third, a master might evaluate a position by assigning that position's successor's successors to slaves, rather than that position's successors. Although this technique involves more message-passing, some advantage might result, because all of a master's slaves would work on finishing the position's first subtree before going on to the second. The evaluation of the second subtree would then receive the full benefit of the beta value generated by the first subtree. Furthermore, when slaves become idle as one subtree is finished, they can immediately be set to work on the next subtree.

Since most game-playing programs must make their move within a certain time limit, any speedup in tree search ability will generally be used to search a deeper lookahead tree. If we have an unlimited supply of processors to form into a binary tree, we can obtain an unlimited speedup only if the search is not limited in time. Otherwise we cannot, because we would eventually violate our premise that the lookahead tree is at least as deep as the processor tree. A new layer on the processor tree does not buy another full ply in the lookahead tree. For example, several speedups of 1.5 would be needed to search a 6-times larger chess lookahead tree, or about one additional ply. The depth of the processor tree would grow faster than the depth of the tree it searches and eventually would catch up. The only way to avoid this limit is to increase the fan-out of the processor tree. If the fan-out is high enough that no successor need ever be queued for evaluation by a slave, then the size of the maximum lookahead tree that can be evaluated within the time limit is limited only by the time required for EVALUATE commands to propagate from the root to the leaves. Long before this limitation is reached, we would run out of silicon for making the processors.

4.7. Analysis of Speedup

We now turn to a formal analysis of the speedup that can be gained in searching large lookahead trees as the number of available processors grows without bound. For this purpose we introduce Palphabeta, a simplified version of the tree-splitting algorithm. This algorithm is in general less efficient than the version already discussed, but is more amenable to analysis. Much of the analysis in this section is a "parallelization" of results of Knuth [38]. Indeed, when $q = 0$ and $f = 1$, Theorem 1 and Corollary 1 reduce to Knuth's results.

As before, the processors will be arranged in a uniform tree. Let $f \geq 1$ be the fan-out of the processor tree (uniform for all interior nodes), and let $q \geq 1$ be its depth (uniform for all terminal nodes). let $q + s$ be the depth of the lookahead tree, where $s \geq 1$. We assume that the lookahead tree has a uniform degree and that this degree, df, is a multiple of f, where d is ≥ 2. Here is Palphabeta:

```
1  function Palphabeta(p:position; α,β:integer) : integer;
2   var i : integer;
3   function g : integer;
4   begin
```

```
5     determine the successors p₁,..., p_df.
6     if depth(p₁) < q then
7        g := Palphabeta
8     else g := alphabeta;
9     for i := 1 to d do
10    begin
11       α:=max(α, max_{(i-1)f<j≤i.f} -g(p_j,-β,-α));
12       if α ≥ β then return(α);
13    end;
14    return(α);
15 end;
```

The f calls to function g in line 11 are intended to occur in parallel, activating functions existing on each of the f slaves. Serial α-β search is activated on leaf slaves; Palphabeta is activated on all others. Unlike the tree-splitting algorithm, Palphabeta waits until all slaves finish before assigning additional tasks. However, the two algorithms behave identically when searching either a best-first or worst-first ordered "theoretical" tree of uniform degree and depth. When we restrict ourselves to one of these lookahead trees, we can therefore make conclusions about the behavior of the tree-splitting algorithm by studying Palphabeta.

4.7.1.a. Worst-First Ordering

α-β search produces no cutoffs if, whenever the call alphabeta(p,α,β) is made, the following relation holds among the successors $p_1, \ldots, p_d$:

$$\alpha < -\text{negamax}(p_1) < \ldots < -\text{negamax}(p_d) < \beta.$$

We call this ordering *worst first*. If no cutoffs occur, it is easy to calculate the time necessary for Palphabeta to finish. Assume that a processor can generate f successors, send messages to all of its f slaves and receive replies in time ρ. (This figure counts message overhead time but does not include computation time at the slaves.) Assume also that the serial α-β algorithm takes time n to search a lookahead tree with n terminal positions. Let a_n be the time necessary for a processor at distance n from the leaves to evaluate its assigned position. A leaf processor executes the serial algorithm to depth s. Thus we have $a_0 = (df)^s$. An interior processor gives d batches of assignments to its slaves, and each batch takes time ρ plus the time for the slave processor to complete its calculation. Thus we have $a_{n+1} = d(\rho + a_n)$. The solution to this recurrence relation is

$$a_q = \rho \left(\frac{d^{q+1} - d}{d - 1}\right) + d^{q+s}f^s,$$

which is the total time for Palphabeta to complete. Since the time for the serial algorithm to examine the same tree is $(df)^{q+s}$, the speedup for large s is f^q. There are $(f^{q+1}-1)/(f-1)$ processors, roughly f^q, so when no pruning occurs the parallel algorithm yields speedup that is roughly equal to the number of processors used.

4.7.2. Best-First Ordering

We will now investigate what happens when the lookahead tree is ordered best-first.

Definition. We will use the Dewey decimal system to name nodes in both processor trees and lookahead trees. The root is named by the null string. The j successors of a node whose name is $a_1 \ldots a_k$ are named by $a_1 \ldots a_k 1$ through $a_1 \ldots a_k j$.

Definition. We say that the successors of a position $a_1 \ldots a_n$ are in *best-first order* if

$$\text{negamax}(a_1 \ldots a_n) = -\text{negamax}(a_1 \ldots a_n 1).$$

Definition. We say a position $a_1 \ldots a_n$ in the lookahead tree is (q,f)-critical if a_i is (q,f)-restricted for all even values of i or for all odd values of i. An entry a_i is (q,f)-restricted if

$1 \leq i \leq q$ and $1 \leq a_i \leq f$
or $q < i$ and $a_i = 1$.

Theorem 4.1. Consider a lookahead tree for which the value of the root position is not $\pm\infty$ and for which the successors of every position are in best-first order. The parallel α–β procedure Palphabeta examines exactly the (q,f)-critical positions of this lookahead tree.

Proof. We will call a (q,f)-critical position $a_1 \ldots a_n$ a *type 1* position if all the a_i are (q,f)-restricted; it is of *type 2* if a_j is its first entry not (q,f)-restricted and n–j is even; otherwise (that is, when n–j is odd), it is of *type 3*. Type 3 nodes have a_n (q,f)-restricted. The following statements can be established by induction on the depth of the position p. (Text in brackets refers to positions of depth $<$ q.)

1) A type 1 position is examined by calling [P]alphabeta($p,+\infty,-\infty$). If it is not terminal, its successor position[s] $p_1[, p_2, \ldots, p_f]$ is [are] of type 1, and $F(p) = -F(p_1) \neq \pm \infty$. This [These] successor position[s] is [are] examined by calling [P]alphabeta($p_i,-\infty,+\infty$). The other successor positions $p_2, \ldots, p_{df}$ [$p_{f+1}, \ldots, p_{df}$] are of type 2, and are all examined by calling [P]alphabeta($p_i,-\infty,F(p_1)$).

2) A type 2 position p is examined by calling [P]alphabeta($p,-\infty,\beta$), where $-\infty < \beta \leq F(p)$. If it is not terminal, its successor[s] $p_1[, p_2, \ldots, p_f]$ is [are] of type 3, and $F(p) = -F(p_1)$. This [These] successor position[s] is [are] examined by calling [P]alphabeta($p_i,-\beta,+\infty$). Since $F(p) = -F(p_1) \geq \beta$, cutoff occurs, and [P]alphabeta does not examine the other successors $p_2, \ldots, p_{df}$ [$p_{f+1}, \ldots, p_{df}$].

3) A type 3 position p is examined by calling [P]alphabeta($p,\alpha,+\infty$) where $F(p) \leq \alpha < +\infty$. If it is not terminal, each of its successors p_i is of type 2, and they are all examined by calling [P]alphabeta($p_i,-\infty,-\alpha$). All of these searches fail high.

It follows by induction on the depth of p that the (q,f)-critical positions, and no others, are examined.

Q.E.D.

Figure 4.2 shows the best-first lookahead tree of degree four and depth four that is examined by Palphabeta running on a processor tree of fanout two and depth two.

Corollary 4.1. If every position on levels $0, 1, \ldots, q+s-1$ of a lookahead tree of depth $q+s$ satisfying the conditions of Theorem 4.1 has exactly df successors, for d some fixed constant, and for f the constant appearing in Palphabeta, then the parallel procedure Palphabeta (along with alphabeta, which it calls), running on a processor tree of fan-out f and height q, examines exactly

$$f^{\lfloor q/2 \rfloor}(df)^{\lceil (q+s)/2 \rceil} + f^{\lceil q/2 \rceil}(df)^{\lfloor (q+s)/2 \rfloor} - f^q$$

terminal positions.

Proof. There are $f^{\lfloor q/2 \rfloor}(df)^{\lceil (q+s)/2 \rceil}$ sequences $a_1 \ldots a_{q+s}$, with $1 \leq a_i \leq df$ for all i, such that a_i is (q,f)-restricted for all even values of i. There are $f^{\lceil q/2 \rceil}(df)^{\lfloor (q+s)/2 \rfloor}$ such sequences with a_i (q,f)-restricted for all odd values of i. We subtract f^q for the sequences $\{1, \ldots, f\}^q 1^s$, that we counted twice.

Q.E.D.

Fig. 4.2. Lookahead tree examined by Palphabeta.

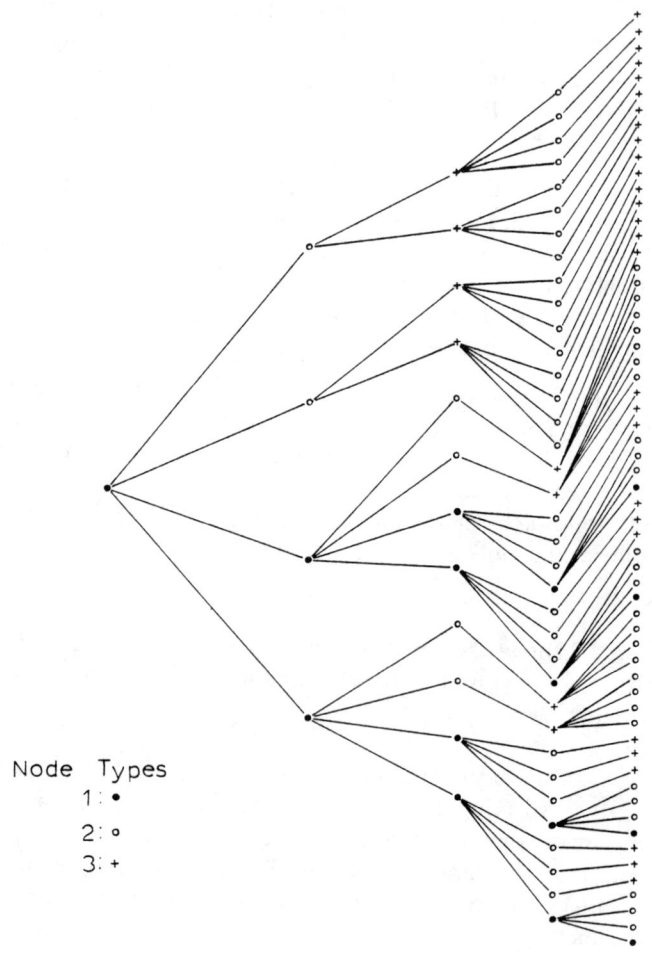

Node Types
1: •
2: ○
3: +

Lemma 4.1. Given positive constants a, b, c, d, and ρ, the relations

$a_0 = a; \quad a_{n+1} = \rho d + a_n + (d-1)b_n;$
$b_0 = b; \quad b_{n+1} = \rho + c_n;$
$c_0 = c; \quad c_{n+1} = d(\rho + b_n).$

are satisfied by the sequences

$$a_n = \begin{cases} \text{(n even:)} \ a + h(n)[d(3\rho+b+c)+\rho-b-c]-n\rho, \\ \text{(n odd:)} \ a + h(n-1)[d(3\rho+b+c)+\rho-b-c]-n\rho \\ \qquad + d^{(n-1)/2}(d(\rho+b)+\rho-b); \end{cases}$$

$$b_n = \begin{cases} \text{(n even:)} \ \rho + 2\rho g(n) + (\rho+b)d^{n/2}, \\ \text{(n odd:)} \ \rho + 2\rho g(n+1) + cd^{(n-1)/2}; \end{cases}$$

$$c_n = \begin{cases} \text{(n even:)} \ 2\rho g(n+2) + cd^{n/2}, \\ \text{(n odd:)} \ 2\rho g(n+1) + (\rho+b)d^{(n+1)/2}; \end{cases}$$

where the function g is defined by

$$g(n) = (d^{n/2} - d)/(d - 1),$$

and the function h is defined by

$$h(n) = (d^{n/2} - 1)/(d - 1).$$

Proof. straightforward algebra.

Theorem 4.2. Under the conditions of Corollary 4.1, and assuming also that 1) serial α-β search is performed in time equal to the number of leaves visited, and 2) in ρ units of time, a processor can generate f successors of a position, send a message to each of its f slaves, and receive the f replies, the total time for Palphabeta to complete is

(q even:) $(df)^{\lfloor s/2 \rfloor} + (df)^{\lceil s/2 \rceil} - 1$
$+ h(q)[d(3\rho+(df)^{\lfloor s/2 \rfloor}+(df)^{\lceil s/2 \rceil}+\rho$
$-(df)^{\lfloor s/2 \rfloor}-(df)^{\lceil s/2 \rceil}] - \rho g,$

(q odd:) $(df)^{\lfloor s/2 \rfloor} + (df)^{\lceil s/2 \rceil} - 1$
$+ h(q-1)[d(3\rho+(df)^{\lfloor s/2 \rfloor}+(df)^{\lceil s/2 \rceil}+\rho$
$-(df)^{\lfloor s/2 \rfloor}-(df)^{\lceil s/2 \rceil}] - \rho g$
$+ d^{(q-1)/2}[d(\rho+(df)^{\lfloor s/2 \rfloor}+\rho-(df)^{\lfloor s/2 \rfloor}).$

Proof. Let a_n, b_n, and c_n represent the time required for a processor at distance n from the leaves of the processor tree to search type 1, 2, and 3 positions, respectively. Then these sequences satisfy the relations

$$a_0 = (df)^{\lfloor s/2 \rfloor} + (df)^{\lceil s/2 \rceil} - 1, \quad a_{n+1} = \rho d + a_n + (d-1)b_n;$$
$$b_0 = (df)^{\lfloor s/2 \rfloor}, \quad b_{n+1} = \rho + c_n;$$
$$c_0 = (df)^{\lceil s/2 \rceil}, \quad c_{n+1} = d(\rho + b_n).$$

By substituting the constant expressions for a_0, b_0, and c_0 to find a_q by the formulas given by Lemma 4.1, we obtain the desired formula.

Q.E.D.

Under conditions of best-first search, the parallel α–β algorithm gives order of $k^{1/2}$ speedup with k processors for searching large lookahead trees. The next theorem formalizes this result:

Theorem 4.3. Suppose that Palphabeta runs on a processor tree of depth $q \geq 1$ and fan-out $f > 1$. Suppose that the lookahead tree to be searched is arranged in best-first order and is of degree df and depth q+s, where $d \geq 1$. Denote by R the time for alphabeta to search this tree, and by P the time for Palphabeta to search the tree. Then

$$\lim_{s \to \infty} R/P = f^{q/2}$$

Proof. The time for the serial algorithm is

$$(df)^{\lfloor (s+q)/2 \rfloor} + (df)^{\lceil (s+q)/2 \rceil} - 1,$$

from Corollary 4.1. If we divide this quantity by the expression given by Theorem 4.2 for P, and take the limit as s goes to ∞, we obtain the desired result.

Q.E.D.

4.7.3. Discussion

The improvement that alphabeta search shows over negamax search is due to the cutoffs it achieves. Parallel execution tends to lose some of that advantage, since subtrees that the serial algorithm would avoid are searched before information is available to cut them off. This situation is most extreme if the lookahead tree is ordered best-first; in this case the serial algorithm enjoys the most cutoffs. However, our analysis shows that even in this case, order of $k^{1/2}$ speedup can still be expected. At the other extreme, if the lookahead tree is ordered worst-first, then no cutoffs are found in either the serial or the parallel algorithm. In this case, the parallel algorithm performs no wasted work, and speedup is order of k.

We can now compare the measurements presented in section 5 with these theoretical bounds. If we take $\gamma_0\gamma_1$ to be the speedup achieved by a processor tree of depth two, then the measured speedup for height-two processor trees of fan-out two and three is 2.64 and 4.59 respectively. Table 3 summarizes theoretical best-first, theoretical worst-first, and measured speedups for processor trees of height one and two, and of fan-out two and three.

Table 3. Speedups

q	f	worst-first	best-first	measured
1	2	2	1.41	1.81
1	3	3	1.73	2.34
2	2	4	2.00	2.64
2	3	9	3.00	4.59

In checkers, certain simplifying assumptions used for the analysis are not true. The lookahead tree is neither regular nor ordered best- (nor worst-) first. Therefore, slave processors do not finish in unison. Nonetheless, our implementation results with checkers display speedups that lie between the two analytically derived extremes. Although tests with more processors should be run, these limited results show that the formal analyses are not unreasonable.

4.7.4. Random Order

Under best-first and worst-first ordering of uniform lookahead trees, sibling slaves finish simultaneously because each slave's pruned lookahead tree has the same size and shape. This fact makes it possible to calculate, for a given processor tree and lookahead tree, the exact finishing time for the algorithm Palphabeta. This section analyzes the behavior of the slightly weaker algorithm Pbound (no deep cutoffs) under the assumption that terminal values are independent, identically distributed random variables. Restated, this assumption says that no two terminal values are equal, and that any one of the n! orderings of the terminal values is as likely as any other. Although the *expected* finishing times for sibling slaves are identical, the finishing times themselves may be unequal. Pbound must therefore wait for the last busy slave to finish before assigning the next batch of tasks. For this reason, we will not attempt to calculate the expected finishing time for the parallel algorithm under conditions of random ordering of terminal nodes. We will, however, present a "parallel" version of Knuth's [38] analysis of the serial algorithm under conditions of random order. The analyses of the parallel and serial cases both yield estimates of the expected number of terminal positions examined. Only in the serial case, however, does this estimate yield a direct estimate of the finishing time of the algorithm.

Here is parallel $\alpha-\beta$ search without deep cutoffs:

function Pbound(p : *position* ; limit : *integer*) : *integer* ;
var m,i,t,d : *integer* ;
begin
   determine the successors $p_1, \ldots, p_{df}$;
   m := $-\infty$;
   if depth(p_1) < q *then* fn := Pbound *else* fn := bound;
   for i := 1 *to* d *do*
   begin t := $\max_{(i-1)f<j\leq i \cdot f}$ $-$fn(p_j,$-$m);
     if t > m *then* m := t;
     if m $\geq$ limit *then* *return*(m);
   end;
   return(m);
end;

Pbound is called with limit = ∞ on the root node of the lookahead tree. On leaf processors, Pbound activates the serial algorithm without deep cutoffs:

function bound(p : *position* ; limit : *integer*) : *integer* ;
var m,i,t,d, : *integer* ;
begin
   determine the successors $p_1, \ldots, p_d$;
   if d = 0 *then* *return*(staticvalue(p)) *else*
   begin m := $-\infty$;
   for i := 1 *to* d *do*
   begin t := $-$ bound(p_i,$-$m);
     if t > m *then* m := t;
     if m $\geq$ limit *then* *return*(m);
   end;
   return(m);
   end;
end;

Let T(d,h) be the number of terminal positions examined by bound(p,∞) in a tree rooted at p of depth h and degree d with randomly distributed terminal values. Knuth [38] establishes that T(d,h) satisfies

$$c_1(d)r_1^h \leq T(d,h) \leq c_2(d)r_2^h,$$

where c_1 and c_2 depend on d but not h, and r_1 and r_2 satisfy $c_3 d/\ln d \leq r_1$ and $r_2 \leq c_4 d/\ln d$, for certain constants c_3 and c_4. As part of the proof of this result, the inequality

(4.1) $\quad (\sum_{1 \leq i \leq d} (\sum_{1 \leq j \leq d} i^{-t((j-1)/2d)})^{s/t})^{1/s} \leq c_4 d/\ln d$

is established for a certain choice of s, t satisfying $1/s + 1/t = 1$.

We begin by presenting a lemma due to Knuth and then adapting it to our own use.

Lemma 4.2. Suppose that $Y_{1,1}, \ldots, Y_{i-1,d}$ and $Z_1, \ldots, Z_{j-1}$ are independent sequences of (i-1)d and (j-1) independent identically distributed random variables. Then

$$\frac{1}{\binom{i-1+(j-1)/d}{i-1}}$$

is the probability that

(4.2) $\quad \max_{1 \leq k < i}(\min(Y_{k,1}, \ldots, Y_{k,d})) < \min_{1 \leq k < j} Z_k$

Proof. If i = 1, the left hand side is $-\infty$. If j = 1, the right hand side is $+\infty$. In both cases, the probability that the relation holds is 1.

Assume then that i,j > 1. Consider the minimum element Y_{k_1,t_1}, over all $1 \leq k_1 < i$ and $1 \leq t_1 \leq d$. The probability that it is less than $\min_{1 \leq k < j} Z_k$ is

$$\frac{(i-1)d}{((i-1)d + j-1)}.$$

Removing the elements $Y_{k_1,1}, \ldots, Y_{k_1,d}$ from consideration, we consider the minimum of the remaining Ys on the left of (4.2), say Y_{k_2,t_2}. The probability that Y_{k_2,t_2} is less than the right-hand side of (4.2) is

$$\frac{(i-2)d}{((i-2)d + j-1)},$$

and so on. Hence (4.2) happens exactly when $Y_{k_1,t_1} <$ RHS and $Y_{k_2,t_2} <$ RHS and ... and $Y_{k_{i-1},t_{i-1}} <$ RHS, so (4.2) has probability

$$\frac{(i-1)d(i-2)d \ldots 1d}{((i-1)d+j-1)((i-2)d+j-1)\ldots(d+j-1)}$$

$$= \frac{(i-1)!((j-1)/d)!}{(i-1+(j-1)/d)!}$$

$$= \frac{1}{\binom{i-1+(j-1)/d}{i-1}}$$

Q.E.D.

Lemma 4.3. (Corollary to Lemma 4.2): If $Y_{1,1}, \ldots, Y_{(i-1)f,df}$ and $Z_1, \ldots, Z_{(j-1)f}$ are independent sequences of $((i-1)f)df$ and $(j-1)f$ independent identically distributed random variables, then the probability p_{ij} that

$$\max_{1 \leq k \leq (i-1)f} \min_{1 \leq m \leq df} Y_{k,m} < \min_{1 \leq k \leq (j-1)f} Z_k$$

is

$$p_{i,j} = \frac{1}{\binom{(i-1)f + (j-1)/d}{(i-1)f}}$$

Proof. This lemma is simply Lemma 4.2 with a change of variables.

Substitute: df for d,
$(i-1)f + 1$ for i,
$(j-1)f + 1$ for j.

Q.E.D.

Since the simple formula k^x is always within 12% of

$$\binom{k-1+x}{k-1},$$

for $0 \leq x \leq 1$ and k a positive integer [38], we will approximate p_{ij} by

(4.3) $p_{ij} = ((i-1)f + 1)^{-(j-1)/d}$.

Theorem 4.4. Let T(d,f,h) be the expected number of terminal positions examined by the parallel α-β procedure without deep cutoffs on a processor tree of degree f and height h in a random uniform lookahead tree of degree df and height h. Then

$$T(d,f,h) < f^h\, c(d,f)\, r(d,f)^h,$$

where r(d,f) is the largest eigenvalue of the matrix

$$M_{d,f} = \begin{pmatrix} \sqrt{p_{11}} & \sqrt{p_{12}} & \cdots & \sqrt{p_{1d}} \\ \sqrt{p_{21}} & \sqrt{p_{22}} & \cdots & \sqrt{p_{2d}} \\ \vdots & \vdots & & \vdots \\ \sqrt{p_{d1}} & \sqrt{p_{d2}} & \cdots & \sqrt{p_{dd}} \end{pmatrix}$$

and c(d,f) is a constant. The quantities p_{ij} in $M_{d,f}$ were defined in Lemma 4.3.

Proof. As before, assign Dewey decimal names to the positions of the lookahead tree. Define the functions

$$G(n) = \lfloor (n-1)/f \rfloor + 1$$

and

$$H(n) = \lfloor (n-1)/f \rfloor f + 1.$$

The nth successor position is a member of the G(n)th batch of successor positions to be assigned to slaves. The first member of that batch is the H(n)th successor position.

When Pbound examines position $a_1 \ldots a_{m-1}$, "limit" is

$$\min_{1 \leq k < H(a_{m-1})} \mathrm{negamax}(a_1 \ldots a_{m-2} k),$$

so its successor $a_1 \ldots a_m$ is examined if and only if $a_1 \ldots a_{m-1}$ is examined and

$$-\min_{1 \leq k < H(a_m)} \mathrm{negamax}(a_1 \ldots a_{m-1} k)$$

$$< \min_{1 \leq k < H(a_{m-1})} \mathrm{negamax}(a_1 \ldots a_{m-2} k)$$

Abbreviate this inequality by P_m. Then $a_1 \ldots a_h$ is examined if and only if $P_1, P_2, \ldots,$ and P_h hold. P_m holds with probability p_{ij}, where $i = G(a_{m-1})$ and $j = G(a_m)$. Furthermore, P_m is a function of the terminal values

staticvalue($a_1 \ldots a_{m-2}jkb_{m+1} \ldots b_n$)
for
$\quad 1 \leq j < H(a_{m-1})$ and all $0 \leq k, b \leq df$
or
$\quad j = H(a_{m-1})$ and $1 \leq k < H(a_m)$ and all $0 \leq b \leq df$.

Therefore P_m is independent of $P_1, \ldots, P_{m-2}$. Let x be the probability that $a_1 \ldots a_h$ is examined. Then we have (assuming, without loss of generality, that h is odd)

$$x < p_{G(a_1)G(a_2)} p_{G(a_3)G(a_4)} \cdots p_{G(a_{h-2})G(a_{h-1})}$$

and

$$x < p_{G(a_2)G(a_3)} p_{G(a_4)G(a_5)} \cdots p_{G(a_{h-1})G(a_h)}.$$

Thus

$$x < \sqrt{p_{G(a_1)G(a_2)}} \sqrt{p_{G(a_2)G(a_3)}} \cdots \sqrt{p_{G(a_{h-1})G(a_h)}}$$

(for even or odd h).
Hence the expected number of terminal positions examined is less than

$$\sum_{1 \leq a_1, \ldots, a_h \leq df} \sqrt{p_{G(a_1)G(a_2)}} \sqrt{p_{G(a_2)G(a_3)}} \cdots \sqrt{p_{G(a_{h-1})G(a_h)}}$$

$$= f^h \sum_{1 \leq a_1, \ldots, a_h \leq d} \sqrt{p_{a_1 a_2}} \sqrt{p_{a_2 a_3}} \cdots \sqrt{p_{a_{h-1} a_h}}$$

$$= f^h \sum_{1 \leq a_1 \leq d} \sum_{1 \leq a_2 \leq d} \sqrt{p_{a_1 a_2}} \sum_{1 \leq a_3 \leq d} \sqrt{p_{a_2 a_3}} \cdots \sum_{1 \leq a_h \leq d} \sqrt{p_{a_{h-1} a_h}},$$

which is $f^h c_{1,h}$, where the sequences $c_{i,n}$, $1 \leq i \leq d$, are defined by

$\quad c_{i,0} = 1$ for $1 \leq i \leq d$

(4.4) $\quad c_{i,n+1} = \sum_{1 \leq j \leq d} \sqrt{p_{ij} c_{j,n}}$, for $1 \leq i \leq d$.

Now define generating functions C_i, for $1 \leq i \leq d$, as follows:

$$C_i(z) = \sum_{n \geq 0} c_{i,n} z^n$$

Then (4.4) is equivalent to

$$C_i(z) - 1 = \sum_{i \leq j \leq d} \sqrt{p_{ij}} z C_j(z) \text{ for } i \leq i \leq d.$$

Set $C(z) = (C_1(z) \ldots C_d(z))^T$, and define the matrix

$$Z = \begin{pmatrix} z\sqrt{p_{11}} & z\sqrt{p_{12}} & \cdots & z\sqrt{p_{1d}} \\ z\sqrt{p_{21}} & z\sqrt{p_{22}} & \cdots & z\sqrt{p_{2d}} \\ \vdots & \vdots & & \vdots \\ z\sqrt{p_{d1}} & z\sqrt{p_{d2}} & \cdots & z\sqrt{p_{dd}} - 1 \end{pmatrix}$$

Then $(-1\ -1\ \ldots\ -1)^T = (Z-I)C$, where I is the identity matrix. By Cramer's rule, $C_1(z) = U(z)/V(z)$, where U and V are polynomials defined by

$$U(z) = \det \begin{pmatrix} -1 & z\sqrt{p_{12}} & \cdots & z\sqrt{p_{1d}} \\ -1 & z\sqrt{p_{22}}-1 & \cdots & z\sqrt{p_{2d}} \\ \vdots & \vdots & & \vdots \\ -1 & z\sqrt{p_{d2}} & \cdots & z\sqrt{p_{dd}} \end{pmatrix}$$

and $V(z) = \det(Z - I)$.

Note that r is an eigenvalue of $M_{d,f}$ if and only if $1/r$ is a root of $V(z)$. Since $C_1(z)$ is a quotient of polynomials, it can be represented [40] as

$$C_i(z) = \sum_{1 \leq k \leq n} G_k(1/(z-B_k)),$$

where $B_1, \ldots, B_n$ are the distinct roots of V, and $G_1, \ldots, G_n$ are polynomials such that the degree of G_i is the multiplicity of B_i.

Every matrix of real, positive elements posseses one positive eigenvalue of multiplicity one that is strictly larger, in absolute value, than all the other eigenvalues [41]. $M_{d,f}$ is positive; let r_i, $i = 1, \ldots, n$, be its eigenvalues, with r_1 the largest. If the eigenvalues $r_1 = 1/B_1, \ldots, r_n = 1/B_n$ of $M_{d,f}$ are distinct, we have

$$C_1(z) = E + \sum_{1 \leq i \leq d} e_i/(z - 1r_i) = E + \sum -e_i r_i/(1 - zr_i),$$

$$= E + \sum_{n \geq 0} \sum_{1 \leq i \leq d} -e_i r_i r_i^n z^n.$$

Since r_1 is the largest of the r_i, $c_{1,h} = O(r_1^h)$. If the eigenvalues of $M_{d,f}$ are not distinct, the representation of $C_1(z)$ involves polynomials of degree higher than one. Even so, the linear term containing r_1 still dominates.

(Q.E.D.)

Lemma 4.4. Suppose the real-valued sequence $a_1, a_2, a_3, \ldots$ obeys the rule

$$a_{m+n} \leq a_m + a_n \quad m,n = 1,2,3,\ldots$$

Then the sequence $a_1/1, a_2/2, a_3/3, \ldots$ either diverges to $-\infty$ or converges.

Proof. It suffices to consider the case where the lim inf, α, of the second sequence is finite. Let $\epsilon > 0$, and choose m such that $a_m/m < \alpha + \epsilon$. Since every integer n can be expressed as $n = qm + r$ with $0 \leq r < m$, we have

$$a_n = a_{qm+r} \leq qa_m + a_r.$$

hence

$$\frac{a_n}{n} = \frac{a_{qm+r}}{qm+r} \leq \frac{qa_m + a_r}{qm+r} = \frac{a_m}{m} \cdot \frac{qm}{qm+r} + \frac{a_r}{n}$$

hence

$$\frac{a_n}{n} < (\alpha+\epsilon) \frac{qm}{qm+r} + \frac{a_r}{n}$$

hence $\lim\sup_{n \to +\infty} a_n/n = \alpha$

and so $\lim_{n \to +\infty} a_n/n = \alpha$.

Q.E.D.

Definition. Let T(d,h) be the number of terminal positions examined by a given algorithm in a lookahead tree of degree d and height h. The *branching factor* of T is

$$\lim_{h \to \infty} T(d,h)^{1/h},$$

if the limit exists.

Theorem 4.5. Let T(d,f,h) be as defined in Theorem 4.4. Then the branching factor of T,

(4.5) $\quad B = \lim_{h \to \infty} T(d,h,f)^{1/h}$,

satisfies

$$\frac{dfc_3}{\log df} \leq B \leq \frac{dfc_4}{\log d}$$

for certain constants $c_3, c_4 > 0$ independent of d and f.

Proof. Since $T(d,h_1,f)T(d,h_2,f)$ is the number of positions that would be examined by Pbound if "limit" were set to $+\infty$ for all positions at height h_1 in a lookahead tree of depth $h_1 + h_2$, we have $T(d,h_1+h_2,f) \leq T(d,h_1,f)T(d,h_2,f)$. Hence by Lemma 4.4 applied to log T(d,h,f), the limit in (4.5) exists.

Lower bound: The parallel α-β routine without deep cutoffs, Pbound, examines at least as many nodes as its serial counterpart, bound, since each "limit" in the parallel case is greater than its counterpart in the serial case. As mentioned above, Knuth has proven that the branching factor of the number of terminal positions examined by bound in a tree of depth h and degree df is greater than or equal to $dfc_3/\log(df)$.

Upper bound: Let s and t be positive real numbers with $1/s + 1/t = 1$, and let E be an eigenvalue of the matrix $A = (a_{ij})$. Suppose $Ax = Ex$. Then

$$|E| \left(\sum_i |x_i^s| \right)^{1/s} = \left(\sum_i \left| \sum_j a_{ij} x_j \right|^s \right)^{1/s}$$

$$\leq \left(\sum_i \left(\sum_j |a_{ij}^t| \right)^{s/t} \right)^{1/s} \left(\sum_j |x_j^s| \right)^{1/s},$$

by Holder's inequality;

hence

$$|E| \leq \left(\sum_i \left(\sum_j |a_{ij}^t| \right)^{s/t} \right)^{1/s}.$$

We will use this inequality to show that $r(d,f) \leq c_4 d / \log d$, for a certain constant c_4 and for $r(d,f)$ as defined in Theorem 4.4. Let $a_{ij} = \sqrt{p_{ij}}$, $E = r(d,f)$, and use approximation (4.3) for p_{ij}. For all s and t such that $1/s + 1/t = 1$, we have

$$r(d,f) \leq \left(\sum_{1 \leq i \leq d} \left(\sum_{1 \leq j \leq d} ((i-1)f+1)^{-t((j-1)/2d)} \right)^{s/t} \right)^{1/s}$$

$$\leq \left(\sum_{1 \leq i \leq d} \left(\sum_{1 \leq j \leq d} i^{-t((j-1)/2d)} \right)^{s/t} \right)^{1/s}$$

$\leq c_4 d / \ln d$, for a suitable s, t, and c_4, by (4.1).

Theorem 4.4 and this upper bound for $r(d,f)$ give us the desired upper bound on the branching factor.

Q.E.D.

Theorem 4.5 deals with lookahead trees that are the same depth as the processor tree that searches them. In the next theorem we extend the analysis to the more general situation in which the lookahead tree can be deeper than the processor tree.

Theorem 4.6. The expected number of terminal positions examined by Pbound in a random uniform game tree of degree df and height q+s, evaluated by a processor tree of degree d and height q, where $d \geq 2$, $q \geq 0$ and $f \geq 1$, is asymptotically less than

$c_5(d,f) \, f^q \, r(d,f)^q \, r_1(df)^s,$

where r(d,f) was given upper and lower bounds in Theorem 4.5, and r_1 satisfies

$$\frac{dfc_3}{\log(df)} \leq r_1(df) \leq \frac{dfc_4}{\log(df)}$$

for the constants c_3 and c_4 appearing in Theorem 4.5, and where $c_5(d,f)$ is a constant independent of q and s.

Proof. Since the values of the positions assigned for evaluation to leaf processors have random values, Theorem 4.4 implies that the number of these positions P satisfies

$$P < c(d,f) \, f^q \, r(d,f)^q.$$

Theorem 4.5 tells us that r(d,f) satisfies

$$\frac{dc_3}{\log(df)} \leq r(d,f) \leq \frac{dc_4}{\log d}$$

If we set "limit" at level q of the lookahead tree to $+\infty$, then each leaf processor evaluating one position at level q would examine less than $c_2(df) r_1(df)^s$ terminal positions [38], where $r_1(df)$ satisfies

$$\frac{dfc_3}{\log(df)} \leq r_1(df) \leq \frac{dfc_4}{\log(df)}$$

and $c_2(df)$ is a constant independent of s.

The result follows with $c_5(d,f)$ set to $c(d,f)c_2(df)$.

Q.E.D.

4.7.5. Discussion of Theorem 4.6.

In searching a lookahead tree of degree df and height q + s, the serial algorithm examines, on the average, at least

$$c_1 \left(\frac{dfc_3}{\log(df)}\right)^{q+s}$$

terminal nodes, where c_1 depends only on df and c_3 is a constant. The parallel algorithm examines less than

(4.6) $\quad c_5 f^q \left(\dfrac{dc_4}{\log d}\right)^q \left(\dfrac{dfc_4}{\log(df)}\right)^q$

terminal nodes on the average.

Under best-first and worst-first ordering, the finishing time for Palpha-beta can be accurately estimated by dividing the amount of work to be done by the number of workers (terminal processors). This method of estimation is somewhat optimistic when applied to Pbound or the Tree-Splitting Algorithm under random ordering, because in Pbound a master waits until all successors in a batch of f have been evaluated before assigning the next batch, and in both Pbound and the Tree-Splitting Algorithm a master waits until the last successor is evaluated before receiving another position.

While we await more powerful methods, let us make the estimate anyway. Dividing (4.6) by the number of terminal processors, f^q, gives us

$$c_5 \left(\dfrac{dc_4}{\log d}\right)^q \left(\dfrac{dfc_4}{\log(df)}\right)^s$$

as the finishing time, and so the speedup would be at least

$$\dfrac{c_1}{c_5} \left(\dfrac{c_3}{c_4}\right)^{s+q} f^q \left(\dfrac{\log d}{\log(df)}\right)^q$$

The factor $(c_3/c_4)^{s+q}$ appears in this expression because we used an optimistic bound for the serial algorithm and a pessimistic bound for the parallel algorithm. We can most likely remove it. The resulting expression is of order

$$\left(\dfrac{f \log d}{\log d + \log f}\right)^q$$

Recall that speedup under worst-first ordering is of order

$$f^q,$$

and by Theorem 4.3, speedup under best-first ordering is of order

$$f^{q/2}.$$

Speedup under both random and best-first ordering is clearly less than speedup under worst-first ordering. Speedup under random ordering is asymptotically greater than speedup under best-first ordering whenever

$$\frac{f \log d}{\log d + \log f} > \sqrt{f};$$

i.e., whenever

$$d > f^{\left(\frac{1}{\sqrt{f}-1}\right)}.$$

4.8. Mandatory-Work-First Search

Under best-first ordering of the lookahead tree, Palphabeta achieves only order of $k^{1/2}$ speedup with k processors. The cause of this inefficiency is clear: As Palphabeta evaluates group after group of children, if one of the children in the group is sufficiently good to produce a cutoff, then all of the work performed on its younger siblings within that group is wasted. If the tree is ordered best-first, all slaves but the first perform needless work. The serial algorithm, under the same ordering, avoids searching these younger siblings.

This section investigates an approach to avoid this extra work in a parallel alpha-beta algorithm. This approach exploits the "mandatory-work-first" distinction first proposed by Akl, Barnard and Doran [37]. By using this distinction to explicitly schedule node evaluations within a tree of processors, we produce a distributed algorithm whose finishing time can be calculated for a given regular processor tree and a given best-first or worst-first lookahead tree. This calculation will show that speedup obtained is "almost optimal" in the number of processors used, in a sense that we will make clear later.

The algorithm, which we will call 'mwf' as short for "mandatory-work-first," is a parallelization of the serial alpha-beta algorithm without deep cutoffs. We briefly review that algorithm, called bound:

>   *function* bound(p : position ; limit : *integer*) : *integer* ;
>   var m,i,t,d, : *integer* ;
>   begin
>   determine the successors $p_1, \ldots, p_d$;
>   *if* d = 0 *then return*(staticvalue(p)) *else*
>   begin m := $-\infty$;

Parallel Alpha-Beta Search

```
        for i := 1 to d do
          begin t := - bound(pᵢ,-m);
            if t > m then m := t;
            if m ≥ limit then return(m);
          end;
        return(m);
      end;
    end;
```

Bound misses some cutoffs achieved by alphabeta, but not many. Under best-first ordering, the tree searched by bound(p,∞) can be described as follows (fig. 4.3): All nodes searched are either type 1 or type 2. The root node is type 1. The first child of a type-1 node is type 1; the remaining children are type 2. The first child of a type-2 node is type 1; the remaining children are cut off. As Knuth and Moore show [38], the branching factor of the tree just described is

$$(d - 3/4)^{1/2} + 1/2,$$

while the branching factor of the alphabeta tree is

$$d^{1/2}$$

under the same conditions of best-first ordering and uniform degree d.

We now define mwf; under best-first ordering, it will examine the same nodes as "bound." As in the serial algorithm, mwf evaluates a type-1 node by recursively evaluating all of its children. Type-2 nodes are only partially evaluated; if necessary, they are later completely reevaluated. In evaluating

Fig. 4.3. Tree searched by bound.

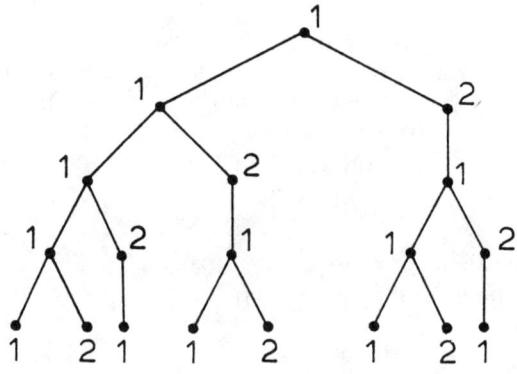

X, a type-2 node, mwf will tentatively predict that its second and later children will be cut off. Only the first child of X is evaluated (completely, since it is type 1), providing a lower bound for the negamax value of X. Later, when the evaluation of Y, the oldest (therefore type-1) sibling of X, is completed, Y's value is compared with X's lower bound. If Y's value is not higher than X's lower bound then the prediction was accurate, and the other children of X need not be evaluated. If Y's value is higher than X's, then X, which has been partially evaluated, must be re-evaluated. X's evaluation is resumed where it left off, and its remaining children are evaluated until all are evaluated or a cutoff occurs.

There are several ways to map this algorithm onto a tree architecture. We will choose a fairly straightforward implementation and suggest modifications later. We will content ourselves with an algorithm that is accurate, if not quick, when moves are not ordered best-first. (If we only insist that the algorithm be accurate when moves are ordered best-first, we could cheat by always picking the first move.) In particular, mwf will simply use the routine intended for evaluation of type-1 nodes when re-evaluating type-2 nodes, even though more sophisticated re-evaluations are possible.

The algorithm uses three functions. "Mwf1(p)" is called on the root node and other type-1 nodes. The position p is to be evaluated. "Mwf2(p)" is called on type-2 nodes. The position p is to be partially evaluated. "Mwf1(p)" is called on positions p that need to be re-evaluated. "Alphabeta" is the serial alpha-beta algorithm with deep cut-offs. It is used by terminal processors to evaluate nodes assigned to them. We assume a processor tree of height q and fanout f. The lookahead tree is of height q+s and degree d. The processor descriptors 1 through d denote the d slave processors. If p is a processor descriptor, then "p.fn()" denotes a remote call of the function "—fn" on processor p. Here is the mandatory-work-first algorithm:

```
1  function mwf1(position p) : integer ;
2  var i,d : integer ;
3      t : array[1..f] of integer;
4      j : processor ;
5  begin
6      if I am a leaf processor then
7          return(alphabeta(p,-∞,+∞));
8      determine the successors p₁, ..., p_d;
9      parfor i := 1 to d do
10         when a slave j is idle do
11             if i = 1 then t[i] := - j.mwf1(p_i);
12             else     t[i] := - j.mwf2(p_i);
13     mwf1 := t[1];
```

```
14      parfor := 2 to d do { re-evaluate if needed }
15      begin
16          when a slave j is idle do
17              if t[i] > mwfl then
18              begin
19                  t[i] := - j.mwfl(p_i);
20                  begincrit
21                      if t[i] > mwfl then mwfl := t[i];
22                  endcrit;
23              end;
24      end;
25  end;
```

Several constructs in mwfl need explanation: First, the construct *parfor* (lines 9 to 12 and 14 to 24) denotes a parallel for-loop. Conceptually, a separate process is created for each iteration of the loop. After all of the processes have completed their iteration, the program continues as a single process at the next statement after the *parfor* loop. Second, the construct "*when* <CONDITION> *do* <BODY>" (lines 10, 17) is similar to "await" in conditional critical regions [42]. The process pauses before <BODY>, proceeding only when <CONDITION> becomes true. This combination of blocking and parallel for-loop implements a queue of positions with the slaves as servers. Third, the construct "*begincrit* <STMT.LIST> *endcrit*" denotes a critical region. Only one process is allowed inside the critical region at a time. The use of a critical region (lines 20 to 22) ensures that the comparison and assignment of mwfl is an atomic operation. The function mwfl calls mwf2 remotely (line 12). Here is mwf2:

function mwf2(p : position) : *integer* ;
p1 : position;
begin
    generate the first successor position p1;
    mwf2 := - mwf1(p_1);
end;

4.8.1. Best-First Order

We now analyze the finishing time of mwf under best-first ordering of the lookahead tree. Define a(i,j) to be the finishing time of mwf in evaluating a type-1 node of height j (in the lookahead tree) on a processor tree of height i and fanout f. An interior processor evaluates a type-1 node by assigning the node's children to its slaves. The d-1 type-2 children are partially evaluated on these slaves by evaluating their type-1 children. If performed by one slave, these evaluations would therefore take time

$$a(i-1,j-1) + (d-1)a(i-1,j-2),$$

not counting message-passing time. With f slaves to do the work, the best time would cut this figure by a factor of f. The worst time occurs if with one type-2 position left in the queue, all f slaves finish their current task simultaneously, and only one can be assigned the final task. The finishing times would be

$$\frac{a(i-1,j-1) + (d-1)a(i-1,j-2)}{f} + m$$

in the best case and

$$\frac{a(i-1,j-1) + (d+f-2)a(i-1,j-2)}{f} + m'$$

in the worst case, where m and m' denote message-passing times. Although m and m' depend on d and f, they are independent of i and j. At the terminal processors, a type-1 node is evaluated with serial alpha-beta search with deep cutoffs. Hence

$$a(0,j) = 2d^{j/2}.$$

To solve this two-dimensional recurrence relation, we need the following lemma. We omit the proof, which involves straightforward algebraic manipulations.

Lemma 4.5. Suppose the two-dimensional sequence a(i,j) obeys the recurrence relation

$$a(i,j) = M(a(i-1,j-1) + Na(i-1,j-2)) + K,$$

and that

$$a(0,j) = Ad^{j/2},$$

where M, N, K, and A are positive real numbers. Then

$$a(i,j) = AM^i d^{(j-i)/2}(1 + Nd^{-1/2})^i + K\frac{(M+MN)^i - 1}{M+MN-1}.$$

This lemma allows us to prove the following theorem.

Theorem 4.7. Suppose that mwf runs on a processor tree of depth $q \geq 1$ and fanout f. Suppose that the lookahead tree to be searched is arranged in best-first order and is of degree $d \geq 2$ and depth q+s, where $s \geq q$. Let X be the finishing time of mwf. Then

$$x \geq 2(1/f)^q d^{s/2}(1+(d-1)d^{-1/2})^q + mf\frac{(d/f)^q - 1}{d-f}$$

and

$$x \leq 2(1/f)^q d^{s/2}(1+d^{1/2})^q + m'f\frac{(d/f)^q - 1}{d-f}$$

Proof. For the first inequality, substitute

1/f for M,
d–1 for N,
m for K,
2 for A,
q+s for j, and
q for i

in the formula given by Lemma 4.5 for a(i,j). For the second inequality, substitute

1/f for M,
d+f–2 for N,
m' for K,
2 for A,
q+s for j, and
q for i

in the formula given by Lemma 4.5 for a(i,j).
                                                                     Q.E.D.

Now that we have calculated the finishing time of the parallel algorithm, we can express the speedup in terms of the number of terminal processors.

Corollary 4.2. Under the assumptions of Theorem 4.7, and under the additional assumption that $s \gg q$, the speedup S of the mwf algorithm with P terminal processors satisfies

$$P^{1-\ln_f(1+d^{-1/2}+(f-2)d^{-1})} \leq S \leq P^{1-\ln_f(1+d^{-1/2}-d^{-1})}$$

Proof. For the lookahead tree under consideration, Corollary 4.1 (Section 4.7.2) says that the finishing time for the serial alpha-beta algorithm with deep cutoffs is approximately

$$2d^{(s+q)/2}.$$

If we divide this quantity by the bounds given on the finishing time given by Theorem 4.7 (Section 4.8.1), and take the limit as s goes to ∞, we obtain the desired result.

Q.E.D.

As d increases, the speedup approaches P, the number of terminal processors. Hence we use the term "almost optimal" to describe mwf under best-first ordering.

4.8.2. Worst-First Order

We defined worst-first order earlier as a particularly poor ordering of the tree under which the serial alpha-beta algorithm achieves no cutoffs. We now analyze the finishing time of mwf1 under the assumption that the lookahead tree is arranged in worst-first order. Mwf1 running on a processor P evaluates a node N by queuing N's children for evaluation on P's slaves. N's first child is evaluated with mwf1, and the others are partially evaluated with mwf2. When partial evaluations are finished, mwf1 discovers it must re-evaluate each of N's d-1 younger children. These children are queued for re-evaluation by mwf1 on P's slaves. In all, we have d invocations of mwf1 and d-1 invocations of mwf2. As in the best-first ordering of the lookahead tree, the f slaves can finish most quickly by finishing their last assignments simultaneously, or most slowly by finishing simultaneously with one more task to be performed. The finishing times would be

$$\frac{da(i-1,j-1) + (d-1)a(i-1,j-2)}{f} + m$$

in the best case and

$$\frac{(d+f-1)a(i-1,j-1) + (d-1)a(i-1,j-2)}{f} + m'$$

in the worst case. A terminal processor evaluates its assigned nodes with the serial alphabeta algorithm. We assume that the serial algorithm evaluates a tree in time equal to the number of leaf nodes on the tree. Hence $a(0,j) = d^j$. With these recursive relationships, we can use Lemma 4.5 to calculate bounds on the finishing times and speedups for mwf1 under worst-first ordering.

Theorem 4.8. Suppose that mwf runs on a processor tree of depth $q \geq 1$ and fanout f. Suppose that the lookahead tree to be searched is arranged in worst-first order and is of degree $d \geq 2$ and depth q+s, where $s \geq q$. Denote by X the finishing time of mwf. Then

$$X \geq (d/f)^q d^s (1 + (d-1)/d^2)^q + mf\frac{(d/f)^q - 1}{d-f}$$

and

$$X \leq ((d+f-1)/f)^q d^s (1+(d-1)/((d+f-1)d))^q + m'f\frac{(d/f)^q - 1}{d-f}$$

Proof. For the first inequality, substitute

d/f for M,
(d−1)/d for N,
m for K,
1 for A,
d^2 for d,
q+s for j, and
q for i

in the formula given by Lemma 4.5 for a(i,j). For the second inequality, substitute

(d+f−1)/f for M,
(d−1)/(d+f−1) for N,
m' for K,
1 for A,

d^2 for d,
q+s for j, and
q for i

in the formula given by Lemma 4.5 for a(i,j).

Q.E.D.

Corollary 4.3. Under the assumptions of Theorem 4.8, and under the additional assumption that $s \gg q$, the speedup S of the mwf algorithm with P terminal processors satisfies

$$P^{1-\ln_f(1+fd^{-1}-d^{-2})} \leq S \leq P^{1-\ln_f(1+d^{-1}-d^{-2})}$$

Proof. For the lookahead tree under consideration, the finishing time for the serial alpha-beta algorithm with deep cutoffs is the number of terminal nodes, which is

d^{s+q}.

If we divide this quantity by the bounds given on the finishing time given by Theorem 4.8, and take the limit as s goes to ∞, we obtain the desired result.

Q.E.D.

4.8.3. Other Orderings

Mwf is deficient in ways that the analysis above does not reveal. First, the re-evaluation of a partially evaluated node searches all the node's children, even though the first child has already been evaluated. A more sophisticated algorithm would resume the search at the second child. This deficiency does not appear under best-first order because no nodes are re-evaluated, and does not significantly affect the algorithm's performance under worst-first order because the re-evaluation involves d times as much work as the partial evaluation. Second, mwf does not attempt to pass $\alpha-\beta$ values to recursive calls on itself, even when these windows are available. This deficiency is insignificant under best-first order because mwf "predicts" all shallow cutoffs that such values could produce. Under worst-first order, cutoffs are not possible and so windows are useless.

These deficiencies occur in the murky area between best-first and worst-first ordering of uniform lookahead trees. The only other "benchmark" lookahead tree available for theoretical treatment is the tree of uniform depth and height with randomly distributed terminal values. But as we have already seen, analyses assuming random ordering are fairly difficult. Further research in this area might be directed toward creating other analyzable orderings of lookahead trees. For example, one might consider a lookahead tree that is originally randomly ordered. As it is being searched, however, heuristics are applied that successfully reorder best branches first in a certain percentage of cases. By making this percentage a parameter, an analysis might be able to model practical situations.

Mwf is a parallelization of the serial algorithm without deep cutoffs. A parallelization of the serial algorithm with deep cutoffs might be possible. Such a parallel algorithm would likely be more complicated than mwf, but might be more efficient.

4.9. Comparison of Palphabeta and mwf

We have now analyzed two different parallel alpha-beta algorithms, Palphabeta and mwf, under conditions of best-first and worst-first ordering. In each of the four possible (algorithm, ordering) combinations, we have derived a formula representing the speedup gained with P terminal processors. Table 4 summarizes these formulas.

Table 4. Speedup in Parallel Alpha-Beta Search

	Ordering of Lookahead Tree	
	Best-First	Worst-First
Palphabeta:	$P^{1/2}$	P
mwf:		
upper bound	$P^{1-\ln_f(1+d^{-1/2}-d^{-1})}$	$P^{1-\ln_f(1+d^{-1}-d^{-2})}$
lower bound	$P^{1-\ln_f(1+d^{-1/2}+(f-2)d^{-1})}$	$P^{1-\ln_f(1+fd^{-1}-d^{-2})}$

The speedups for mwf depend on d, the degree of the lookahead tree, and f, the fanout of the processor tree. It is instructive to substitute actual values for d and f. For example, the average number of moves from a position in the game of chess is about 38. For a best-first lookahead tree of degree 38 and processor tree of fanout 2, Corollary 4.2 predicts that speedup for mwf will satisfy

$$P^{0.78} \leq S \leq P^{0.82},$$

which is significantly better than Palphabeta. For a worst-first lookahead tree of degree 38 and processor tree of fanout 2, Corollary 4.3 predicts that speedup for mwf will satisfy

$P^{0.93} \leq S \leq P^{0.96}$,

which is almost as good as Palphabeta.

4.10. Tips for Processor-Tree Architects

Anyone designing hardware to play a game such as chess faces a number of decisions along the way. Our discussion raises the following questions:

1) Should parallel processing be used?
2) If so, how powerful should the leaf processors be?
3) How many leaf processors should be used?

Since our algorithms all reduce to the serial algorithm when q=0, the third question is really a generalization of the first. To help answer these questions, we make the following simplifying assumptions:

1) A certain fixed amount of money may be spent.
2) The parallel algorithm gives P^e speedup with P processors for some fixed $0 < e \leq 1$.
3) Several different serial processors are available. Associated with each of these processors is a dollar cost and a processing speed in units of positions examined per second. These data are described by a "serial power function," W(S), that tells how much power W we can obtain in a serial processor by spending S dollars. We will assume that W(S) is continuously differentiable.

4.10.1. Serial versus Parallel

We will start with a concrete example. Suppose that for $40,000 we can buy a processor that searches 250 nodes per second, and for $10,000 we can buy a processor that searches 100 nodes per second. Suppose that we have $40,000 to spend and a parallel algorithm that gives $P^{0.5}$ speedup with P processors. We can spend our money to buy one $40,000 processor or four $10,000 processors. We know that our algorithm will give $4^{0.5}$ speedup, or 200 nodes per second with the four processors. Hence we would be wise to buy the $40,000 processor and use the serial algorithm to evaluate 250 instead of 200 nodes per second. In general, when we multiply the amount of money available to us by any constant k, we do better to use serial processing if we gain more than k^e speedup with the more expensive serial processor.

We define the *critical point*, if it exists, to be that number of dollars S such that optimal systems less expensive than S are serial machines, and optimal systems more expensive than S are parallel machines. If W(S) follows the classic economic pattern of diminishing returns on investment for large S (fig. 4.4), then the critical point occurs when

$$\frac{d\log W}{d\log S} = e.$$

The following theorem helps to formualize this result.

Theorem 4.9. Suppose that W is a positive, differentiable function defined on the positive real numbers. Suppose that e > 0, and that there exists a positive number S_0 such that for $S \geq S_0$,

$$\frac{d\log W}{d\log S} \leq e.$$

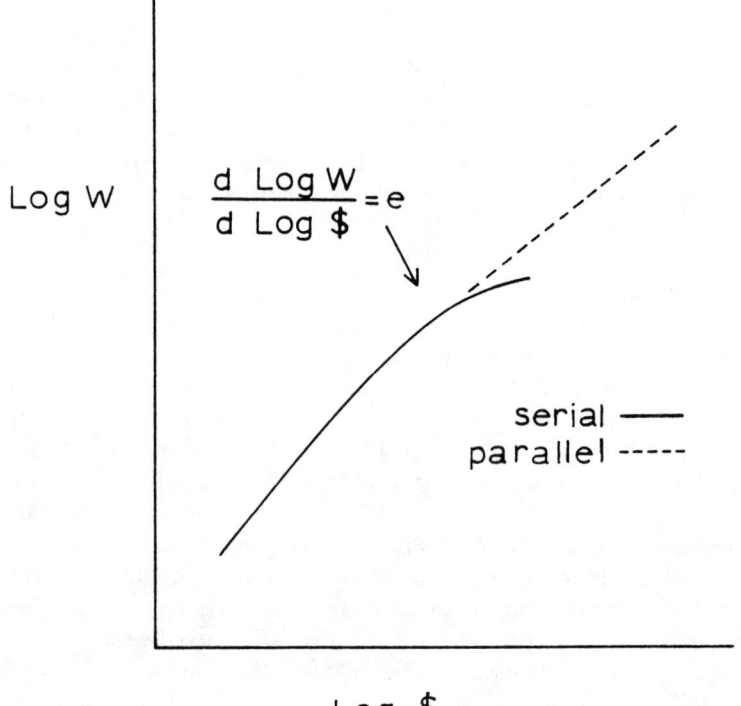

Fig. 4.4. Serial vs. parallel machines.

Then for $k > 1$,

$$k^e W(S_0) \geq W(kS_0).$$

Proof. The proof is by contradiction. Suppose that

$$k^e W(S_0) < W(kS_0).$$

Taking the log of both sides, we can rewrite this as

(4.7) $\quad e < \dfrac{\log W(kS_0) - \log W(S_0)}{\log kS_0 - \log S_0}.$

By the Mean Value Theorem, the right-hand side of 4.7 is equal to the derivative

$$\frac{d\log W}{d\log S}$$

evaluated at some point S_1 such that $S_0 \leq S_1 \leq kS_0$. Hence

$$e < \frac{d\log W}{d\log S} \ .$$

at S_1, which contradicts our original assumption.

Q.E.D.

Theorem 4.9 says that if the marginal return on investment in serial machines after S_0 dollars is less than $e\%$ improvement in speed for every 1% increase in cost, then more speed can be obtained from k processors at S_0 dollars apiece than from a single machine at kS_0 dollars.

4.10.2. *Maximal Processor Trees*

Our analyses all assume that each message to a leaf processor invokes a substantial amount of work. For the moment, let us assume that a "substantial amount of work" consists of a search of a subtree of height one or more. As we have pointed out, when move selection must be completed with a certain time limit, we cannot gain unlimited speedup by adding more and more layers to the processor tree. Each additional layer in the processor tree buys less than one additional layer in the lookahead tree that can be searched within the time limit. Hence the subtrees assigned to terminal processors grow shorter and shorter, and eventually we violate our assumption about giving them substantial amounts of work per message. In this section we estimate how many processors are required before the assumption becomes false.

Suppose that a serial processor can search a lookahead tree L of depth D within the required time limit. Assume that the branching factor (the ratio of the sizes of lookahead trees of successive depths when searched by the serial algorithm) is B, and that the fanout of the processor tree is f. Assume that the parallel algorithm we are using gives k^e speedup with k leaf processors, for some fixed $0 < e \leq 1$. A processor tree of height q gives f^{eq} speedup. Within the time limit it can therefore search a pruned lookahead tree that is f^{eq} times as big as L, or $\log_B f^{eq}$ ply deeper. Hence Palphabeta assigns subtrees of depth one to leaf processors when

$$D + \log_B f^{eq} = q + 1,$$

or when

(4.8) $$q = \frac{D-1}{1-e\log_B f}.$$

When the processor tree is shorter than the q given by Equation 4.8, our advertised speedups can be met within the time limit. Hence for $P < f^q$ leaf processors, we obtain speedup P^e. Mwf assigns subtrees of height one to leaf processors when the height of the processor tree is one less than half the height of the lookahead tree. Hence for mwf, the equation corresponding to (4.8) is

$$q = \frac{D-1}{2-e\log_B f}.$$

5
Piecewise-Serial Iterative Methods

Beyond the Cray 2, a yet faster computer is taking shape in Mr. Cray's mind. "I do tend to look forward in my thinking and I don't like to rest on my laurels," he says. How fast could such a computer be? Perhaps, he says, a trillion calculations a second.

That prospect is an intriguing one for scientists. Says Sidney Fernbach, a scientific administrator at Livermore, "There's no machine that Seymour Cray can conceive that would be too fast for us."

Wall Street Journal
(12 April 1979)

5.1. Introduction

Many numerical computations are *locally defined* and *iterative*: A rectangular array of numbers A_0 is given; $A_1, A_2, \ldots$ are iteratively defined by a *locally defined rule*. That is, the value of an element in A_n is some function of the values of its immediate neighbors in A_{n-1}.

This chapter investigates locally defined iterative computations on Arachne-like architectures. As a focus for our investigation we will consider the numerical solution of an important problem in engineering and physics, the *Dirichlet problem*.

Section 2 discusses the numerical solution of the Dirichlet problem. Section 3 discusses previous work in parallel iterative methods. Section 4 proposes a family of distributed algorithms for the iterative solution of the Dirichlet problem.

5.2. The Dirichlet Problem

Let R be the interior and S the boundary of the unit square $0 \leq x \leq 1, 0 \leq y \leq 1$. Let g(x,y) be a continuous function defined on S. In the Dirichlet problem we seek a function u(x,y) defined on R + S that is twice continuously differentiable on R and satisfies *Laplace's equation*

56 Piecewise-Serial Interative Methods

(5.1) $u_{xx} + u_{yy} = 0$

on R, and equals g(x,y) on S. To approximate u(x,y) we superimpose over R + S a uniform mesh of N+1 horizontal and N+1 vertical lines with spacing h = 1/N, for some positive integer N. We call the $(N+1)^2$ intersections of these lines *mesh points*. To approximate u at a given internal mesh point (x,y), we use the approximations

$$u_{xx} \cong [u(x+h,y) + u(x-h,y) - 2u(x,y)]/.h^2$$
$$u_{yy} \cong [u(x,y+h(+ u(x,y-h) - 2u(x,y)]/h^2$$

to rewrite (5.1) as

(5.2) $4u(x,y) - u(x+h,y) - u(x-h,y) - u(x,y+h) - u(x,y-h) = 0.$

This equation applied at interior points together with the boundary condition

$$u(x,y) = g(x,y)$$

forms a discrete version of the Dirichlet problem.

5.2.1. Jacobi Method

Equation (5.2) specifies a set of $(N-1)^2$ linear equations in $(N-1)^2$ unknowns. This set of equations could be solved directly, but the sparsity of the matrix often makes iterative methods more efficient. We first solve equation (5.2) for u(x,y), giving

$$u(x,y) = [u(x+h,y) + u(x-h,y) + u(x,y+h) + u(x,y-h)]/4.$$

Given "old" values $u_n(x,y)$ at mesh points, we use the following equation to generate "new" values $u_{n+1}(x,y)$:

(5.3) $u_{n+1}) = [u_n(x+h,y) + u_n(x-h,y) + u_n(x,y+h) + u_n\ (x,y-h)]/4.$

Equation (5.3) is applied iteratively until further iterations do not change u very much. This method is the *Jacobi* (J) method.

The Jacobi method is very slow for large N. Indeed, it is well known [43] that the number of iterations required for the Jacobi method to converge is proportional to N^2. The total work needed is proportional to N^4, since each iteration treats $O(N^2)$ internal mesh points.

Although slow, the Jacobi method is useful for two reasons. First, for many problems its intermediate iterates correspond to the transient behavior of the physical process being modeled. Many other methods converge more quickly to the steady state, but do so by mathematical shortcuts not taken by the physical process being modeled. When we are interested in transient behavior for problems of heat flow, we must use methods similar to the Jacobi method. Second, some optimizations of J such as SOR (defined below) are unstable for some problems other than the Dirichlet problem.

5.2.2. Gauss-Seidel Method

The *Gauss-Seidel* (GS) method differs from the Jacobi method by using new neighbor values whenever available. For example, if the outer loop of each iteration visits rows from y=0 to y=1, and the inner loop visits mesh points in a row from x=0 to x=1, then GS calculates new values according to the formula

$$u_{n+1} = [u_n(x+h,y) + u_{n+1}(x-h,y) + u_n(x,y+h) + u_{n+1}(x,y-h)]/4.$$

The GS method needs only half as many iterations to converge as the J method. This speedup, though significant, is independent of N. Hence GS also needs $O(N^2)$ iterations to achieve convergence.

5.2.3. Successive Over-Relaxation

GS optimizes J by using new values whenever available. *Successive Over-Relaxation* (SOR) optimizes GS by "over-correcting" from one iteration to the next. If GS computes the value u'_{n+1} by adding the increment $u'_{n+1} - u_n$ to u_n, then SOR computes u_{n+1} by adding an even greater increment:

$$u_{n+1} = u_n + w(u'_{n+1} - u_n).$$

The *relaxation parameter* w is usually between 1 and 2; if 1, then SOR reduces to GS. Much work has been done to determine optimum values of w. For the Dirichlet Problem on the unit square with mesh spacing h, it can be shown that the optimum value of w is $2/(1 + \sin(h\pi))$ [43]. For example, if h is 1/20 then the optimal value for w is 1.72945.

With an optimal value of w, SOR requires $O(N)$ iterations to converge.

58 Piecewise-Serial Interative Methods

5.3. Previous Work

This section reviews parallel algorithms that have been developed for locally defined iterative methods.

Stone [28] notes that many serial techniques are not directly applicable in parallel algorithms. For example, the serial Gauss-Seidel technique uses newly computed values for neighboring points wherever possible. If new values for all points are calculated simultaneously, then Gauss-Seidel cannot be used.

Rosenfeld [44] simulates the operation of a C.mmp-like machine (that is, all processors have equal access to all memories) in the computation of the distribution of current in an electrical network. He shows that with proper programming aimed at reducing storage interference, N processors can give nearly N-fold speedup when N is less than about 10.

Weiman [45] proposes an L by M grid of microprocessors to perform iterative calculations on an L by M by N-point mesh arising from the Navier-Stokes equation. Processors are connected in the four compass directions. For a certain range of problem size, each cell needs approximately 2K words of storage, a small number of registers, and a small processor. Cell (i,j) holds all data points with spatial coordinates (i,j,x); after each time step it communicates all newly computed values to all four neighbors.

Flanders [6] has built a 32-by-32 SIMD array of one-bit microprocessors called the Distributed Array Processor (DAP). Communications lines connect each processor to its neighbors in the four cardinal directions. Finite-difference calculations are performed by mapping the processors one-to-one onto the points of the problem grid. If the problem grid is larger than the processor grid, then the calculation for each time step must successively load the processor array with "patches" from the problem grid. Flanders estimates that a 64-by-64 DAP array would perform finite-difference calculations at a rate 20 times that of an IBM 360/195.

Welch [46] reports measurements of calculations used in atmospheric simulation models on the Pepe Parallel Processor, which is an SIMD machine with data transfers on a shared bus. Measurements were taken on Pepe hardware with 11 processing elements (PEs). Extrapolation of these measurements indicates that a 161-PE Pepe would execute the Geophysical Fluid Dynamics Laboratory benchmark about 7 times faster than an IBM 360/195.

Parallel iterative methods usually perform exactly the same computation as some well-known serial method. An interesting exception is the *chaotic relaxation* technique of Chazan and Mirankar [47] and Baudet [48]. Chaotic relaxation is an attempt to avoid time-consuming synchronization among multiprocessors performing iterative methods. For example, suppose several processors access a common memory to perform Jacobi's method. If we insist that exactly the same computation be performed as in the serial case, then when processor x is computing the value of a point whose neighbor p is computed by processor y, the two processors must synchronize their actions so that processor y's computation of p for the nth iteration is stored before processor x accesses p. The overhead of synchronization can significantly slow down the computation. Chaotic relaxation doesn't bother with synchronization; x gets either the old or the new value for p. Theoretical analyses indicate that as long as one processor does not lag too far behind its neighbors, convergence is still assured, if slowed. Baudet's measurements on C.mmp show that the method pays off: The extra iterations needed for convergence are more than offset by the time saved by not synchronizing.

5.4. Piecewise-Serial Iterative Methods

This section analyzes several parallel algorithms for implementing the Jacobi method on a distributed system. As usual, we define *speedup* to be time for the serial algorithm divided by time for the parallel algorithm. The *efficiency* of a parallel algorithm is its speedup divided by the number of processors used.

5.4.1. Uniform Regions With Grid Topology

One natural way to solve the Dirichlet problem on a multicomputer architecture is the following: Arrange the processors in a q by q grid, with each processor connected to its nearest neighbors in the four compass directions. We assume that the problem grid consists of pq by pq points. This grid is broken into q^2 square regions of size p^2 (fig. 5.1). If we index both the processor grid and problem grid by Cartesian coordinates, then processor (i,j), for $0 \leq i,j < q$, contains those problem points (x,y) such that $iq \leq x < (i+1)q$ and $jq \leq y < (j+1)q$. Hence regions sharing a common border are assigned to neighboring processors.

We will call the following algorithm the *grid algorithm*. In order to compute values for the next time step according to the Jacobi method, each processor needs to know the current value of points bordering on its region. Prior to the computation of each time step, each processor communicates to each of its nearest neighbors values of its border points adjacent to that neighbor. We assume that a processor can send n numbers to a neighbor in time $\rho + n\delta$ (ρ is the per-message overhead, and δ is the time to send one number) and can compute one mesh point value in one unit of time. Each processor must send and receive p numbers to/from each of its four neighbors, so the communication phase of the grid algorithm takes

60 Piecewise-Serial Interative Methods

Fig. 5.1. Partition of problem grid for grid topology.

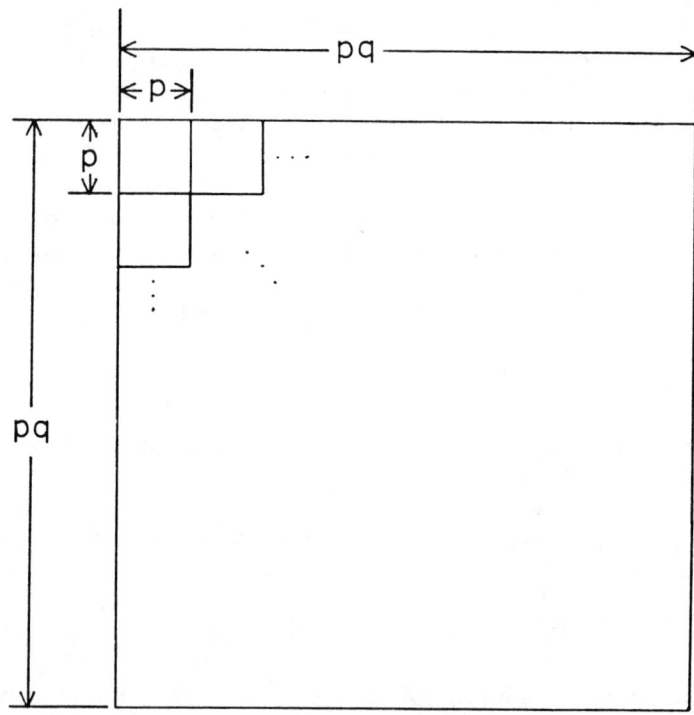

$8\rho + 8p\delta$

units of time. The computation phase takes p^2 units of time to compute p^2 points on each processor. We have proved the following result:

Theorem 5.1. The grid algorithm computes one iteration of the Jacobi method in

(5.4) $8\rho + 8p\delta + p^2$

units of time.

The first two terms of (5.4) represent message-passing time; the last term represents computation time. Speedup is

$$\frac{p^2 q^2}{8\rho + 8p\delta + p^2}$$

and efficiency is

$$\frac{p^2}{8\rho + 8p\delta + p^2}.$$

5.4.2. Uniform Regions with Tree Topology

A more flexible way to implement the Jacobi method on a multicomputer architecture is the *Synchronous Tree Algorithm*: Suppose that the problem grid is pn^q by pn^q, containing p^2n^{2q} points in all. Arrange the processors into a tree of height q and fanout n^2. (The height of a tree with one node is zero.) Any nonterminal processor is called a *master* and its children are called *slaves*.

The root processor is responsible for the entire problem grid. Having n^2 slaves, it divides the grid into n^2 square subregions of size pn^{q-1} by pn^{q-1} and assigns each subregion to a slave. Each of these n^2 slaves likewise divides its region into n^2 regions, assigning each to one of its slaves, and so on. Terminal slaves receive a square region with p^2 mesh points (fig. 5.2).

Before each time step, any slave (terminal or otherwise) needs to know point values from the previous time step that border on its region. Its master provides these values. At the end of each time step, each slave sends to its master all of its border point values. The master will relay appropriate sets of values to its slaves before the next time step.

We now calculate the finishing time for one time step (fig. 5.3). Define a_k to be the time for a processor at distance k from the leaves of the processor tree to complete a time step. We start timing after the processor has gotten its border values from its master. We stop timing when the processor is ready to send border values to its master. Since a slave processor must calculate p^2 values, $a_0 = p^2$. A processor at distance k+1 from the leaves must first supply $4pn^k$ border values to each of its n^2 slaves. Each slave then takes time a_k and sends approximately $4pn^k$ border values ($4pn^k - 4$, to be exact) back to the master. At time t=0, we start timing. At time $t = i(\rho + 4\delta pn^k)$ the ith slave, for $i = 1, \ldots, n^2$, finishes receiving border points and starts computing values for the next iteration. At time $t = i(\rho + 4\delta pn^k) + a_k$ the ith slave finishes its computation phase and starts to send border points to its master. (The master may still be busy sending points to the latter slaves when the first slave wants to send points back to it. For the time being, we assume that if necessary a master can simultaneously send and receive without being slowed down in either activity. Later we will find conditions under which no overlap occurs.) At time $t = n^2(\rho + 4\delta pn^k) + a_k + \rho + 4\delta pn^k$, the last slave finishes sending border points to its master. We therefore have the recurrence relation

(5.5) $\quad a_{k+1} = (n^2+1)(\rho + 4\delta p n^k) + a_k,$

which is the finishing time in the synchronous algorithm for a master at height k+1, in terms of the finishing time of one of its slaves. The solution to this recurrence relation for k = q is

$$a_q = (n^2+1)(q\rho + 4\delta p(n^q-1)/(n-1)) + p^2.$$

We have proved

Theorem 5.2. The finishing time a_q for the synchronous tree algorithm is

(5.6) $\quad a_q = (n^2+1)(q\rho + 4\delta p(n^q-1)/(n-1)) + p^2.$

Hence speedup is

Fig. 5.2. Partition of problem grid for tree topology.

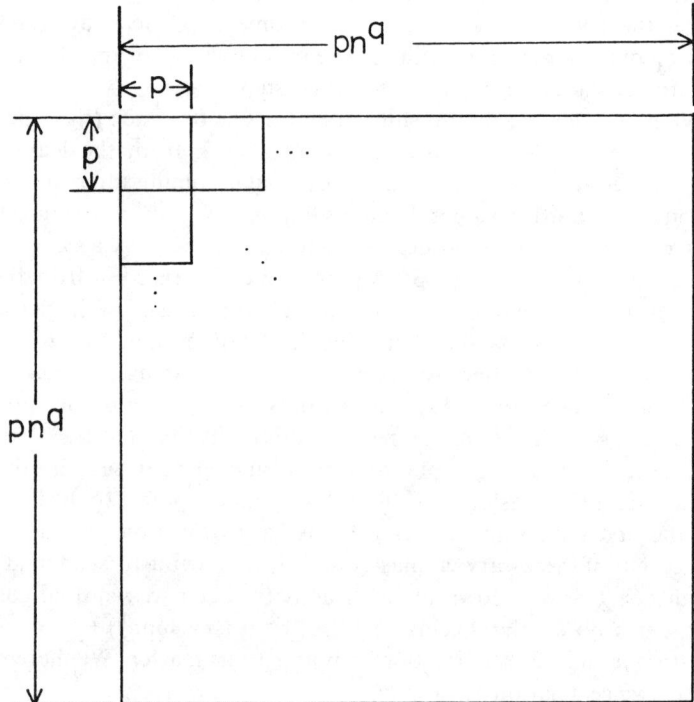

$$\frac{p^2 n^{2q}}{(n^2+1)(q\rho + 4\delta p(n^q-1)/(n-1)) + p^2}$$

and efficiency is

$$\frac{p^2}{(n^2+1)(q\rho + 4\delta p(n^q-1)/(n-1)) + p^2},$$

which is less than the efficiency of the grid algorithm.

Optimal Fanout

We have assumed that the fanout of the processor tree is some perfect square n^2. We now show that the optimal value of n is two. Suppose we have a fixed number of leaf processors C^2. There may be several different pairs of integers

Fig. 5.3. One time step in the life of a master.

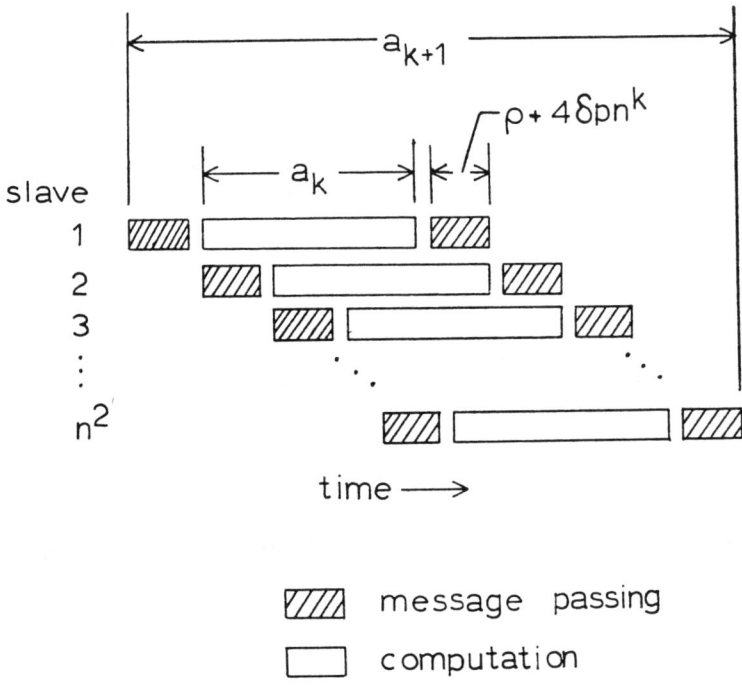

64 Piecewise-Serial Interative Methods

(n,q) such that a tree of height q and fanout n^2 has $C^2 = n^{2q}$ terminal processors. As (5.6) shows, for a given problem the computation time for these several trees is identical (p^2), but the message-passing time may vary. We wish to find the optimal fanout n^2; i.e., we want to know which value $2 \leq n \leq C$ minimizes the message-passing time

$$(n^2+1)(q + 4\delta p(n^q-1)/(n-1)).$$

Theorem 5.3. Let p and C be arbitrary positive integers, and let ρ and δ be positive real numbers. If we let n range over the integers greater than one, and require that $n^q = C$, then

$$(n^2+1)(qp + 4\delta p(n^q-1)/(n-1)).$$

achieves its minimum value at n = 2.

Proof. We show that both

$$(n^2+1)q \quad \text{and} \quad (n^2+1)(n^q-1)/(n-1)$$

achieve their minimum value at n = 2.
 First, $n^q = C$ and q = (ln C)/ln n. To minimize

$$(n^2+1)q = (n^2+1)(\ln C)/\ln n,$$

it would suffice to minimize $g(n) = (n^2+1)/\ln n$, since ln C > 0. Since the derivative

$$g'(n) = \frac{2n(\ln n) - (n^2+1)/n}{\ln^2 n}$$

is always positive for $n \geq 2$, g(n) achieves its minimum at n = 2.
 To minimize

$$\frac{(n^2+1)(n^q-1)}{n-1}.$$

we first note that $n^q-1 = C-1 > 0$, so it would suffice to minimize

$$h(n) = \frac{n^2+1}{n-1}.$$

Again, the derivative h'(n) is positive for n ≥ 3, and h(2) = h(3) = 5, so h achieves its minimum at n = 2.

Q.E.D.

We have shown that 2^2 is the optimal perfect square fanout of the processor tree. This fact might suggest to us that a fanout of 2 might yield even greater efficiency. We now show that this conjecture is not true. Suppose, for example, that we have a tree of height 2q and fanout 2, yielding the same number of leaf processors, 2^{2q}, as a tree of height q and fanout 4. The master M at height 2k+2 divides the $p2^{k+1}$ by $p2^{k+1}$ problem grid into two *rectangular* subregions, each of shape $p2^{k+1}$ by $p2^k$, and assigns each to one of its two slaves, s_1 and s_2. Slave s_n divides the rectangular region into two square regions of side $p2^k$, and assigns each to one of its slaves, s_{n1} and s_{n2}. Define b_k to be the time for a processor at height 2k to finish one time step. We define a_q according to equation (5.5) with n=2.

Theorem 5.4. For a given number of leaf processors, the synchronous tree algorithm finishes sooner on a processor tree of fanout four than on a processor tree of fanout two.

Proof. We develop a recurrence relation for b_k and compare it to the recurrence for a_k. We start our timing when M starts sending border points to s_1. We have the following sequence of events: At t=0, M starts sending $3p2^{k+1}$ points to s_1. At $t=\rho+3\delta p2^{k+1}$, M starts sending $3p2^{k+1}$ points to s_2. At $t=2\rho+6\delta p2^{k+1}$, s_2 has received all points and starts sending $2p2^{k+1}$ points to s_{21}. At $t=3\rho+8\delta p2^{k+1}$, s_2 starts sending $2p2^{k+1}$ points to s_{22}. At $t=4\rho+10\delta p2^{k+1}$, s_{22} finishes receiving points and starts its computation. At $t=4\rho+10\delta p2^{k+1}+b_k$, s_{22} finishes its computation and starts sending $2p2^{k+1}$ (minus 4, which we ignore) points back to s_2. At $t=5\rho+12\delta p2^{k+1}+b_k$, s_2 finishes receiving points from s_{22} and starts sending $3p2^{k+1}$ (minus 4) points back to M. At $t=6\rho+15\delta p2^{k+1}+b_k$, M finishes receiving points from s_2 and is thus finished with its computation. Hence

$$b_{k+1} = 6(\rho + 5\delta p2^k) + b_k,$$

whereas

$$a_{k+1} = 5(\rho + 4\delta p2^k) + a_k.$$

Since $a_0 = b_0 = p^2$, we have $b_q > a_q$, for $q > 0$.

Q.E.D.

66 Piecewise-Serial Interative Methods

Among processor trees, we may now restrict our attention to those whose fanout is four. Equation (5.6), the finishing time for the tree architecture, simplifies to

(5.7) $\quad a_q = 5q\rho + 20\delta p(2^q - 1) + p^2,$

which is the finishing time for a processor tree of height q and fanout four executing the synchronous algorithm.

Nonoverlap of Send and Receive

We now return to the possibility that a slave might wish to send border points to its master before that master has finished giving border points to all the other slaves. Suppose that the master is at height q from the leaves of the processor tree. After the first slave has finished receiving its border points, two activities proceed in parallel (fig. 5.4). First, the first slave performs its computation. This activity takes time

Fig. 5.4. Nonoverlap of send and receive.

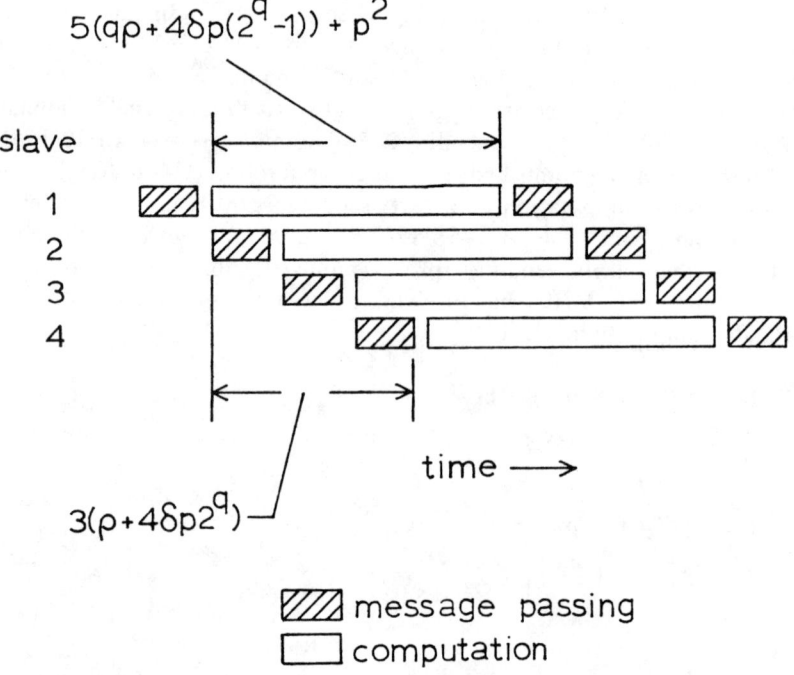

(5.8) $5(q\rho + 4\delta p(2^q - 1)) + p^2,$

according to Equation (5.7). Meanwhile, the master is sending border points to the other three slaves. This activity takes time

(5.9) $3(\rho + 4\delta p 2^q).$

If the computation (5.8) takes longer than the communication (5.9), we can be sure that the master will not need to receive results while it is sending values. We therefore require the condition

$$5(q\rho + 4\delta p(2^q - 1)) + p^2 - 3(\rho + 4\delta p 2^q) > 0.$$

We can rewrite this inequality as

$$\rho(5q-3) + \delta p(2^{q+3} - 20) + p^2 > 0.$$

The left-hand side increases monotonically with q; hence we have the greatest difficulty in satisfying the inequality when q=0. It therefore suffices to show that

$$p^2 - 12\delta p - 3\rho > 0.$$

By the quadratic formula, we have proved

Theorem 5.5. In a tree of arbitrary depth and fanout 4 executing the synchronous algorithm, if the condition

(5.10) $p > 6\delta + \sqrt{36\delta^2 + e\rho}$

is met then no slave will ever be ready to send points to its master before that master is finished sending points to the other slaves.

M-Dimensional Problem Space

The tree architecture can easily be adapted to locally defined iterative methods in M dimensions. We assume that the fanout of the processor tree is some perfect Mth power, n^M. The height of the processor tree is q. We assume that the original problem grid is pn^q points on a side, with $(pn^q)^M$ points in all. Each master divides its region into n^M regions. Thus each terminal processor is assigned a region with p^M points. Using methods similar to those in the two-dimensional case, we can calculate a_q, the finishing time for one step, as

$$a_q = (n^M+1)(q\rho+2M\delta p^{M-1}\;\frac{n^{(M-1)q}-1}{n^{M-1}-1}) + p^M.$$

As in the two-dimensional case, we can prove that for a given number of terminal processors, 2^M is the optimal fanout among perfect Mth powers.

Semi-Synchronous Method

We have seen that for a given number of leaf processors, the tree architecture is less efficient than the grid architecture. The cause of this inefficiency is that the leaf processors sit idle while masters higher in the tree exchange border points. This section investigates a technique for decreasing this inefficiency. The technique, called the *semi-synchronous algorithm*, is based on the observation that a slave need not perform all of the computation for one time step before sending border points to its master. After receiving border points, the slave can compute its border points first, then immediately send them to its master. Thus the communication of border points up and then down the processor tree has a head start and proceeds in parallel with the remaining computation on leaf processors. We want to give conditions under which the batch of border points for the next time step are ready for the slave when that slave is done with the current time step.

From the slave's point of view, a time step starts when border points start arriving from its master. At time $t=(\rho+4\delta p)$, the slave finishes receiving its neighbors' border points and starts computing its own border points. After computing its border points, the slave requests to send them, and goes back to computing the rest of the points. When the master is ready to receive, the slave interrupts its computation and takes $\rho+4\delta p$ units of time to send its 4p border points. After these points are sent, the slave resumes computing interior points. Hence the slave can complete its cycle for the semi-synchronous algorithm in

(5.11) $2(\rho+4\delta p) + p^2$

units of time.

In the semi-synchronous method, we require that leaf slaves rush to send freshly computed border points to their master. The masters, by contrast, will execute the same algorithm as in the synchronous algorithm. We therefore assume that the fanout of the processor tree is four, since we have already seen that this fanout minimizes total exchange time. From the point of view of the master of a terminal slave, a time step starts when it has finished receiving border points from its master and attempts to send border points to its first slave. Since that slave may still be busy with computation from the previous time step, the master must wait a certain amount of time x. At time x, the master starts sending points to the first slave (fig. 5.5). At time x + 4(ρ+4δp), the master starts receiving points from the first slave, and at time x + 8(ρ+4δp) finishes receiving points from all the slaves. We

Fig. 5.5. One time step in the semisynchronous method.

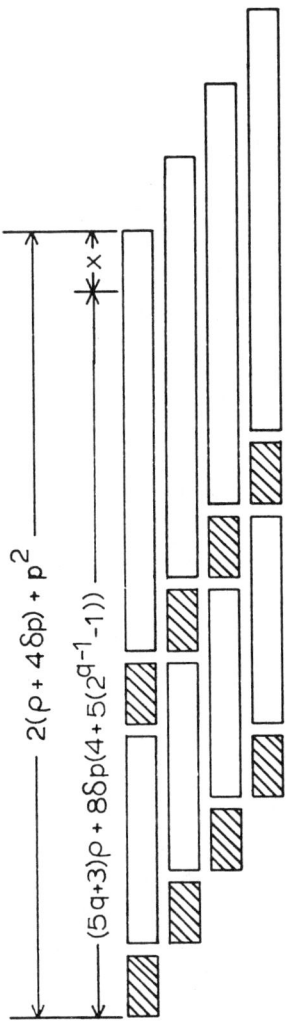

assume that the first slave is ready to start sending points when the master is ready to start receiving them. This assumption is true when the computation time for the border points, 4p, is less than or equal to the time to pass point values to the other three slaves, $3(\rho + 4\delta p)$. The necessary condition is

$$3\rho + 4p(3\delta-1) > 0.$$

Let b_k, for $k \geq 1$, be the time for a master at height k from the leaves to complete one time step. We have just shown that $b_1 = x + 8(\rho+4\delta p)$. Masters above level one see the same behavior in their neighbors as in the synchronous algorithm. Hence

$$b_{k+1} = b_k + 5(\rho+4\delta p 2^k).$$

The solution to this recurrence relation for k=q is

(5.12) $\quad b_q = x + (5q+3)\rho + 8\delta p(4+5(2^{q-1}-1)),$

which is the finishing time of a processor tree of height q and fanout four that is executing the semi-synchronous algorithm, when masters of terminal slaves must wait time x to send points to their slaves.

When $x > 0$, the cycle time of the terminal slaves is greater than the cycle time of the rest of the processor tree, and so b_q also equals (5.11), the cycle time for a slave:

(5.13) $\quad x + (5q+3)\rho + 8\delta p(4+5(2^{q-1}-1)) = 2(\rho+4\delta p) + p^2.$

With equation (5.13), the condition $x > 0$ becomes

$$p^2 - 8\delta p(3+5(2^q-1)) - \rho(5(q-1)+6) > 0.$$

Hence when

(5.14) $\quad p > 4\delta(3+5(2^q-1)) + \sqrt{16(\delta(3+5(2^q-1)))^2 + \rho(5(q-1)+6)},$

the slave processors are never idle, and the slave cycle time (5.11) represents the finishing time of the algorithm. When $x > 0$ we say that the tree is *compute-bound*. When (5.14) is not satisfied, (5.12) with x=0 gives the finishing time, and the tree is *exchange-bound*.

When the semi-synchronous method is compute bound, its finishing time, $2\rho + 8\delta p + p^2$, is less than the finishing time of the grid topology, $8\rho + 8\delta p + p^2$. In the semi-synchronous method a slave can accomplish all of its sending and receiving in two messages, but the grid-topology algorithm requires eight messages, leading to the 6ρ time difference. Since the semi-synchronous method gives the exchange of border points a head start, it always finishes sooner than the synchronous tree method. The grid algorithm, the synchronous tree algorithm and the semi-synchronous method all give nearly n-fold speedup on large problems: The speedup in all three algorithms approaches the number of slave processors as the problem size goes to infinity.

5.4.3. Efficiency

Theorem 5.1 shows that the efficiency e of the grid algorithm is

$$e = \frac{p^2}{8\rho + 8p\delta + p^2}.$$

Theorem 5.2 shows that the efficiency e of the synchronous tree algorithm is

$$e = \frac{p^2}{5(q\rho + 4p\delta(2^q-1)) + p^2}$$

The efficency e of the semi-synchronous tree algorithm is

$$e = \frac{p^2}{2\rho + 8p\delta + p^2}$$

if it is compute-bound. Solving each these three equations for p, and representing $e/(1-e)$ by W, we get

(5.15) $\quad p = 4\delta W + 2\sqrt{4\delta^2 W^2 + 2\rho W}$

for the grid architecture,

(5.16) $\quad p = 10W\delta(2^q-1) + \sqrt{(10W\delta(2^q-1))^2 + 5Wq\rho}$

for the synchronous tree architecture, and

(5.17) $\quad p = 4\delta W + \sqrt{16\delta^2 W^2 + 2\rho W}$

for the compute-bound semi-synchronous tree architecture. For given δ, ρ and q, these equations describe the minimum value of p for which the various algorithms will yield a given efficiency e. As the desired efficiency

72 *Piecewise-Serial Interative Methods*

increases to one we must put a larger and larger subproblem on each leaf or grid processor.

5.4.4. Measurement of Communication Time

We have defined δ to be the ratio of per-point communication time to per-point computation time. In this section we present measurements taken on a VAX-11/780 and a PDP-11/70 that can be used to estimate δ for those machines. The following C subroutine was computed on each machine:

```
#define SIDE 50
double buffer[SIDE][SIDE];
GS(k) int k; {
   register double *p;
   register int j,i;
   double quart = 1/4.0;

   for(i=0;i<SIDE;i++){
      buffer[0][i] = buffer[SIDE-1][i]
      = buffer[i][0] = buffer[i][SIDE-1] = 9.9;
   }
   for(;k>0;k--){
      p = &buffer[1][1];
      for(i=1;i<SIDE-1;i++){
         for(j=1;j<SIDE-1;j++){
            *p=(*(p-SIDE) + *(p+SIDE)
               + *(p-1) + *(p+1))
               * (quart);
            p++;
         }
         p += 2;
      }
   }
}
```

This subroutine implements the Gauss-Seidel method. We have gone to some effort to produce efficient code in this example. For example, we rejected array access by subscript because repeated address calculations produce a program about four times slower than the above. The C compiler on the VAX produced the following assembler code for the nested "i" and "j" loops at the end of the routine:

Piecewise-Serial Interative Methods

```
        movl    $1,r9               /initialize i loop
L5:     movl    $1,r10              /initialize j loop
L3:     addd3   400(r11),-400(r11),r0   /south + north
        addd2   -8(r11),r0          /add west neighbor
        addd2   8(r11),r0           /add east neighbor
        muld3   -12(fp),r0,(r11)    /multiply by 1/4.0
        addl2   $8,r11              /move pointer 1 step east
        aoblss  $49,r10,L3          /j loop control
        addl2   $16,r11             /move pointer to next row
        aoblss  $49,r9,L5           /i loop control
```

The C compiler on the PDP-11/70 produced the following assembler code for the same C code segment:

```
        mov     $1,r3       /initialize i loop
L5:     mov     $1,r2       /initialize j loop
L3:     movf    -620(r4),r0 /add north neighbor
        addf    -10(r4),r0  /add west neighbor
        addf    620(r4), r0 /add south neighbor
        addf    10(r4),r0   /add east neighbor
        mulf    -20(r5),r0  /multiply by 1/4.0
        movf    r0,(r4)     /store result
        add     $10,r4      /move pointer 1 step east
        inc     r2          /j loop control
        cmp     $61,r2
        jgt     L3
        add     $20,r4      /move pointer to next row
        inc     r3          /i loop control
        cmp     $61,r3
        jgt     L5
```

On both the VAX and PDP-11, the routine was executed with an argument of 500, so that the body of the "k" loop was executed 500 times. Since "SIDE" was defined to be 50, the body of the "j" loop was executed 48*48*500 times. Elapsed time on the VAX was 20.3 seconds, on the PDP-11 was 34.0 seconds. Time per point on the VAX is therefore 20.3/48*48*500 = 16.24 microseconds, and on the PDP-11 is 34.0/48*48*500 = 27.2 microseconds.

The DA11-B DMA Unibus link can transfer 500,000 16-bit words per second between two Unibuses. At this rate, the communication time for one 64-bit floating point word is 8 microseconds. If we use this hardware for connecting machines, $\delta_{VAX} = 0.49$ and $\delta_{PDP-11/70} = 0.29$.

We have defined ρ as the ratio of per-message overhead time to per-point computation time. Estimating ρ is harder than estimating δ because ρ depends on the operating system as well as the hardware. One distributed operating system for which we can estimate ρ is Arachne. Assuming 32-bit floating point numbers, computation time per point on an LSI-11 is approximately 500 microseconds. Per-message overhead on Arachne is approximately 12 milliseconds. Hence $\rho_{Arachne}$ = 24. Since time to send one floating-point word, ignoring per-message overhead, is about 0.4 milliseconds, $\delta_{Arachne}$ = 0.8. These sample figures for ρ and δ are not necessarily representative of real distributed systems.

The values of ρ and δ strongly influence the efficiency of distributed systems in solving the Dirichlet problem. We can use these estimates to gain information about various topologies of LSI-11s running the Arachne distributed operating system. For example, these figures along with Theorem 5.5 require p $\geq$ 15 in order to guarantee that sends and receives not overlap in the synchronous algorithm. For the semi-synchronous algorithm, relation (5.14) tells us that for trees of heights one and two, the relations p $\geq$ 54 and p $\geq$ 118 respectively must hold in order that the slave cycle time (5.11) would represent the finishing time. Otherwise the finishing time is represented by equation (5.12) with x set to zero. Relation (5.15) tells us that for the grid architecture to achieve an efficiency of 50%, p $\geq$ 18 must hold. For the grid architecture to achieve an efficiency of 75%, p must be greater than or equal to 36. Relation (5.16) requires p $\geq$ 22 and 53 for the synchronous algorithm running on trees of height one and two, respectively, to achieve an efficiency of 50%.

5.4.5. Scheduling Tree Machines

An entire tree machine need not be devoted to a single problem. For example, a tree of height 3 and fanout 4 can also be considered as 4 trees of height 2, 16 trees of height 1, or 64 trees of height 0 (serial processors). Since we can never achieve n-fold speedup from n leaf processors, the efficiency of a tree machine is less than the efficiency of its workers, the leaf processors. Hence, if we have a large queue of problems, throughput is maximized by using the 4^q leaf processors as individual servers.

But throughput is not likely to be the primary concern of builders and users of parallel architectures. A parallel machine sacrifices low cost and efficiency to gain speed in the execution of lengthy tasks. As hardware becomes less expensive, this tradeoff becomes more profitable. Speed may be desired for various reasons. We may want to meet a time constraint (as in weather prediction or real-time applications), or we might simply want to minimize time spent by a human waiting for a computation. For such a machine to accomplish its purpose, the average interarrival time for tasks must be significantly greater than the average execution time. Otherwise a queue would form, defeating the original purpose, which is speed. Hence the scheduling of a tree machine is not likely to be a complicated issue. To achieve the primary goal of speed, a scheduling algorithm that devotes the entire tree to each task until completion would likely be satisfactory. During those periods when no lengthy tasks are requesting service, the tree-machine may be partitioned into individual processors for other tasks.

5.4.6. Static Nonuniform Regions

As long as the problem grid is uniform and square, the grid architecture is sufficient and the tree architecture unnecessary. Unfortunately, many problems cannot be modeled in such a regular way. First, models of moving fluids often require extra grid points in places where values are fluctuating rapidly. Second, computation is often conditional. For example, a weather model may skip the calculation of some radiation terms at a point if clouds are present. Additional radiation terms may be omitted if it is night at a point. Third, the geometry of the problem space may be irregular. All of these situations create problems for a rectangular grid architecture, whether MIMD as discussed here, or SIMD as in the DAP [6].

The tree topology provides a solution to these problems by allowing load leveling through *region encroachment*. When load is evenly distributed throughout a square region, each master divides its region into four uniform smaller regions and assigns one region to each slave. When load is unevenly distributed, a master can divide its region into four *equal-load* smaller regions in the following way: First, divide the region along one dimension into two strips of equal load. Second, divide each of these strips along their long dimension into two regions of equal load. The four regions, called *skewed quadrants*, receive equal load (fig. 5.6). If each master follows this procedure, terminal slaves receive regions of equal load. Figures 5.7 through 5.10 illustrate this procedure carried out for a tree architecture of fanout four and height two, and for problem regions in the form of a disc, half-disc, annulus, and thin strip. Figure 5.11 illustrates the procedure carried out for a tree architecture of fanout four and height three when the load in the unit square is distributed according to $x^3 + y^3$.

5.4.7. Dynamic Region Encroachment

At the cost of $1/3$ more processors for communications, the tree topology gains the advantage of flexibility over the grid topology. As we have seen, this flexibility allows load leveling by region encroachment among those

Fig. 5.6. Skewed quadrants.

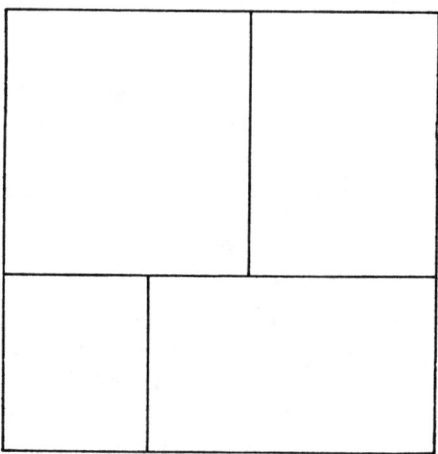

processors actually doing the calculation. If the load function varies with time, this load leveling can occur dynamically. For example, at each time step a master can decide if the load in its region is evenly divided among its four slaves by noting their computation times. Since a slave may spend part of its time rebalancing its own slaves, its master can discover the actual computation time only by being told by the slave. Thus each message bearing border points from a slave to its master should include the computation time for that slave. If a slave is overloaded, the master may decide to equalize load with a new set of skewed quadrants. The master then sends messages to those slaves that will lose points and receives in reply the points that will be transferred. These points are then forwarded to the appropriate slaves, after which the normal sequence of events resumes. Since the load leveling itself entails a cost, a master must weigh that cost against the expected speedup. Load leveling between slaves would occur only when imbalance exceeds a certain threshold.

Since a master can also be a slave, it must also be prepared for messages from its master that give or take away points. A master M follows the following algorithm:

1) Receive message from master.
2) If the message is a request for points, forward it to the affected slaves and assemble their replies into a reply to the master. If the message is a set of points to be added to M's domain, divide the points into one

Fig. 5.7. Disc.

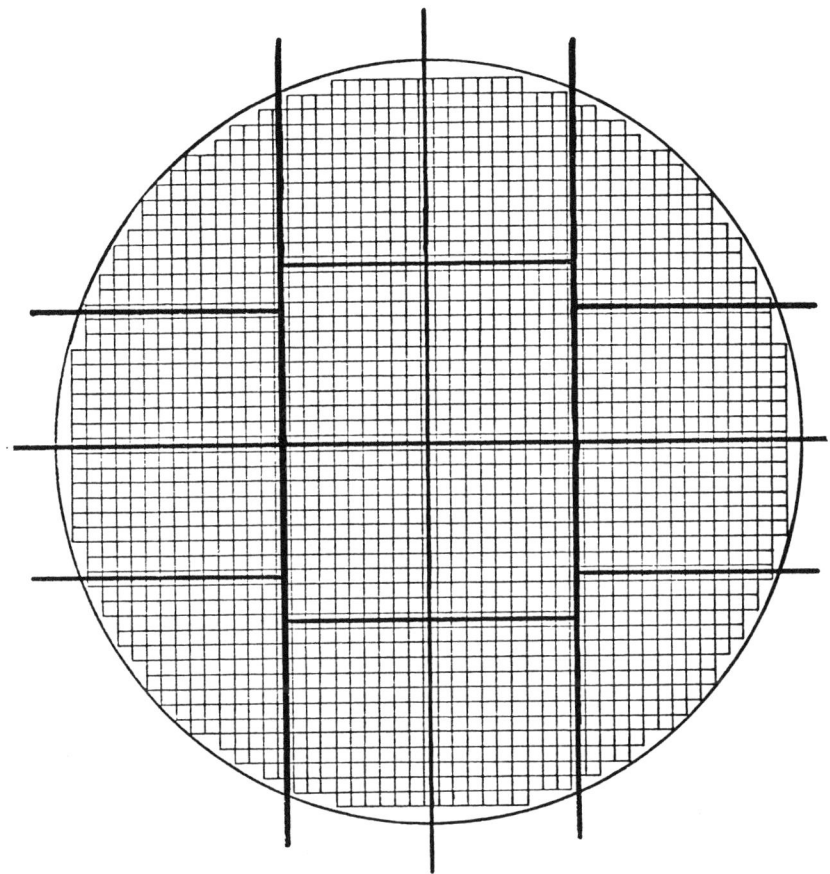

package for each affected slave and relay the packages to these slaves. Go to 1).
3) If the message is the usual update of border points, relay these updates to slaves. Go to 1).
4) Decide whether load leveling needs to be done among the slaves. If load leveling is needed, send messages to slaves that lose points and relay the points in their replies to the appropriate siblings. Go to 1).

Fig. 5.8. Half-disc.

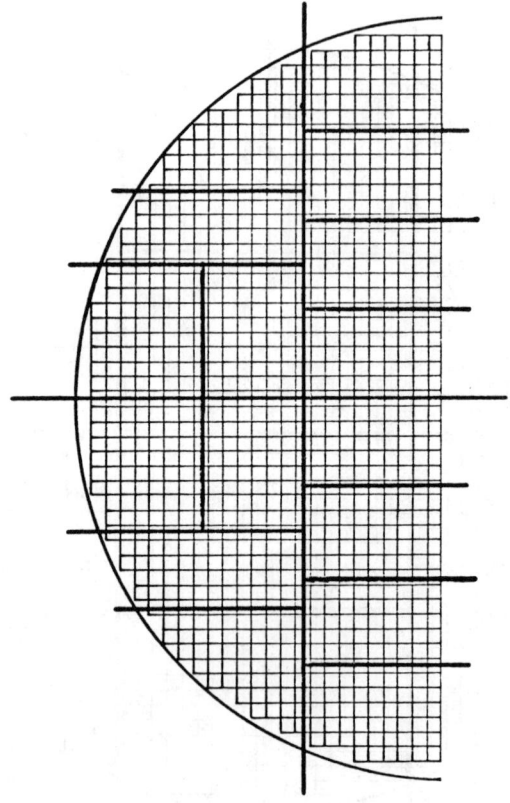

Fig. 5.9. Annulus.

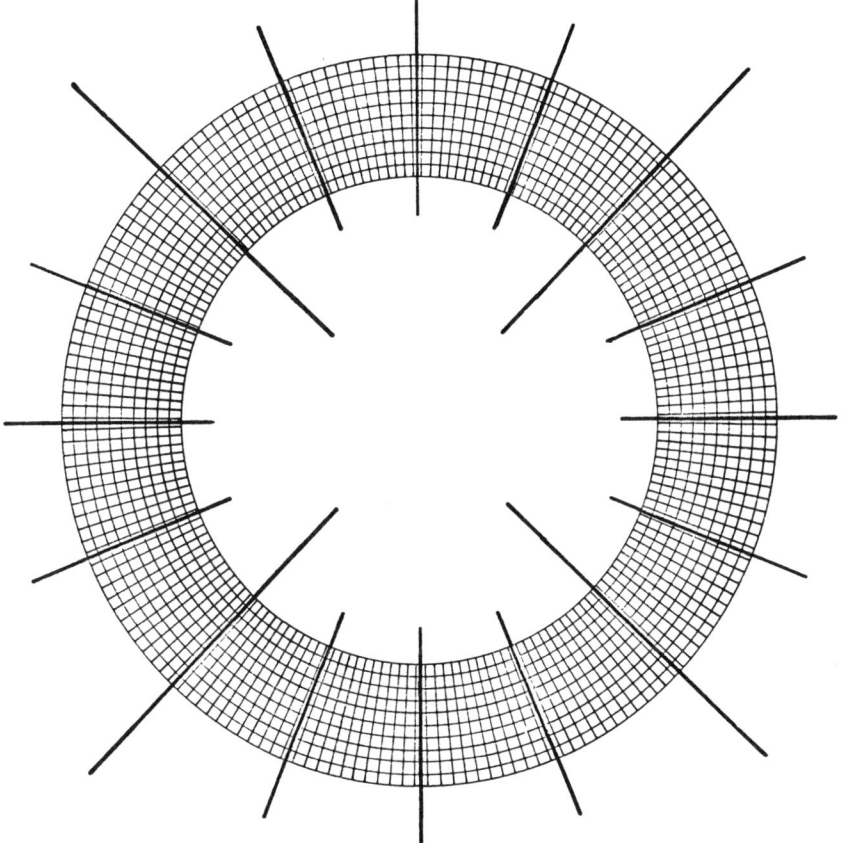

Fig. 5.10. Thin strip.

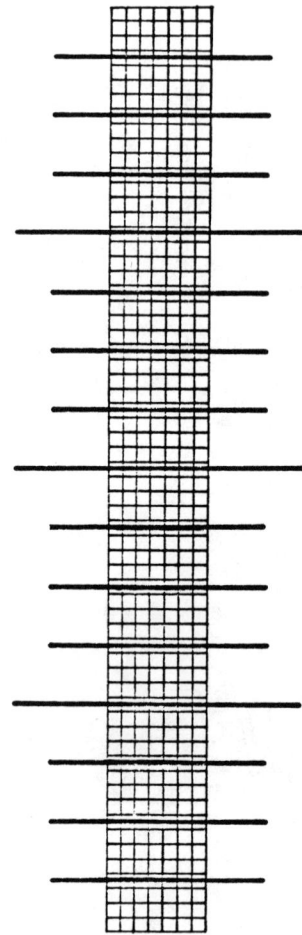

Fig. 5.11. Load = $x_3 + y_3$ on unit square.

82 Piecewise-Serial Interative Methods

Fig. 5.12. Load = bivariate normal distribution with standard deviation a) 100; b) 4; c) 3; d) 2.5.

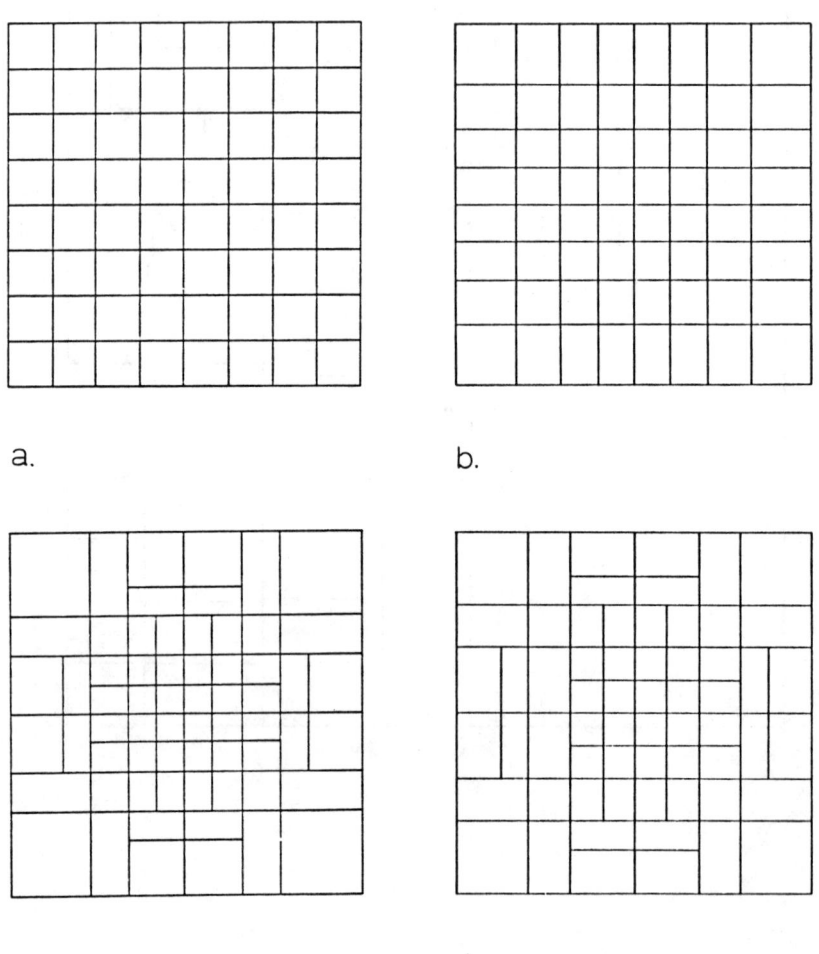

a.

b.

c.

d.

Even if the load does not vary with time, dynamic region encroachment could be used to adapt automatically to an arbitrary uneven load. The regions could initially be as for an even load, and would converge toward a configuration with equal loads.

Figure 5.12 illustrates dynamic region encroachment for a region in which the load is a bivariate normal distribution on the square $(-5,5) \times (-5,5)$ with a time-varying standard deviation.

6

Quotient Networks

It is a profoundly erroneous truism, repeated by all copy-books and by eminent people when they are making speeches, that we should cultivate the habit of thinking what we are doing. The precise opposite is the case. Civilization advances by extending the number of important operations which we can perform without thinking about them.

Alfred North Whitehead quoted in *A Certain World* by W. H. Auden

6.1. Introduction

One barrier to the practical use of interconnection networks is the lack of algorithms for processing large problems on small machines. (By an interconnection network we mean an SIMD parallel computer interconnected by some interconnection strategy.) Often, it is assumed that N processors are available to process N data [13,49]. If we happen to have N+1 or more data points, then we must choose between the serial algorithm and a bigger machine. Examples of interesting exceptions can be found in work by Baudet and Stevenson [21], and by Siegel, Mueller and Siegel [50]. This chapter investigates a method that constructs algorithms for solving large problems on small networks. We call these algorithms *quotient-network algorithms*. Section 2 reviews some proposed interconnection networks. Section 3 reviews proposed algorithms for those networks. We call these algorithms *large-network algorithms*, since each one assumes as many processors as points in the problem to be solved. Section 4 presents a general method for transforming a large-network algorithm into a quotient-network algorithm. Section 5 applies this method to each of the algorithm-machine combinations of section 3. Section 6 discusses some economic advantages of quotient-network algorithms.

6.2. Existing Networks

In this section we briefly review some proposed interconnection networks. For a more thorough overview, see [51]. We assume that each network contains N processors. We denote the square root of N by n, and $\log_2 N$ by m. We will name the processors PE(0) through PE(N–1). Sometimes we refer to a processor by the binary form of its number, $p = p_{m-1}p_{m-2} \cdots p_1p_0$.

6.2.1. Grid-Connected Network

In this network, the processors are arranged in a two-dimensional n by n grid. The processor in the ith row and j^{th} column is named PE(i,j), for $0 \leq i,j < n$. A processor is connected to its north, south, east and west neighbors:

If $i > 0$, PE(i,j) is connected to PE(i–1,j).
If $i > n-1$, PE(i,j) is connected to PE(i+1,j).
If $j > 0$, PE(i,j) is connected to PE(i,j–1).
If $j < n-1$, PE(i,j) is connected to PE(i,j+1).

The Illiac IV network adds additional connections among edge processors [2].

6.2.2. Perfect Shuffle

Shuffle-Exchange

In this network, PE($p_{m-1}p_{m-2} \cdots p_1p_0$) is connected to PE($p_{m-2} \cdots p_1p_0p_{m-1}$) by the "shuffle function" line and to PE($p_{m-1}p_{m-2} \cdots p_1\overline{p_0}$) by the "exchange function" line.

4-Pin Shuffle

In this network, each processor has two input pins IPIN0 and IPIN1, and two output pins OPIN0 and OPIN1. We can number all input pins by assigning to IPIN0 on processor $p_{m-1}p_{m-2} \cdots p_1p_0$ the number $p_{m-1}p_{m-2} \cdots p_1p_00$. IPIN1 on the same processor is assigned the number $p_{m-1}p_{m-2} \cdots p_1p_01$. This numbering allows us to refer to input pins as IPIN(0) through IPIN(2N–1). Output pins are numbered in the same way, OPIN(0) through OPIN(2N–1). The shuffle function is used to transfer data from the output pins of PE($p_{m-1}p_{m-2} \cdots p_1p_0$) to the input pins of processors PE($p_{m-2} \cdots p_1p_00$) and PE($p_{m-2} \cdots p_1p_01$). (The name "4-pin shuffle" is new.)

6.2.3. PM2I

In the Plus-Minus 2^i network (PM2I), PE(j) is connected to processors

$j+2^i \mod N$

and

$$j - 2^i \bmod N,$$

for $0 \leq i < m$.

6.2.4. Cube

$PE(p_{m-1} \cdots p_{i+1} p_i p_{i-1} \cdots p_0)$ in the cube network is connected to the m processors $PE(p_{m-1} \cdots p_{i+1} \overline{p}_i p_{i-1} \cdots p_0)$, for $0 \leq i < m$.

6.3. Existing Algorithms

In this section we review some proposed large-network algorithms. The number of such algorithms is large and growing, so we do not attempt to be comprehensive. Our goal is to illustrate the process of transforming large-network algorithms into quotient-network algorithms.

6.3.1. Fast-Fourier Transform on the Shuffle

Let $A(k)$, $k = 0, 1, \ldots, N-1$, be a vector of N complex numbers. The *Discrete Fourier Transform* (DFT) of A is defined to be the vector

$$(6.1) \quad X(j) = \sum_{k=0}^{N-1} A(k) W^{jk} \quad j = 0, 1, \ldots, N-1$$

where $W = e^{2\pi i/N}$. The obvious algorithm for computing X takes time $O(N^2)$. An important advance in the theory of algorithms was the discovery of an $O(N \log N)$ algorithm for the DFT [24]. This algorithm is called the *Fast Fourier Transform* (FFT). Pease [10] has discovered an algorithm that computes the DFT on $N/2$ processors in time proportional to $\log N$, thus achieving optimal speedup. Pease's algorithm can be explained as follows: First, we represent both k and j by their binary expansion.

$$k = k_{m-1} k_{m-2} \cdots k_0$$

and

$$j = j_{m-1} j_{m-2} \cdots j_0$$

Equation 6.1 then becomes

86   Quotient Networks

(6.2)
$$X(j) = \sum_{k_0} \sum_{k_1} \ldots \sum_{k_{m-1}} A(k_{m-1}k_{m-2}\ldots k_0) W^{jk_{m-1}2^{m-1}} W^{jk_{m-2}2^{m-2}} \ldots W^{jk_0}$$

$$= \sum_{k_0} W^{jk_0} \sum_{k_1} W^{jk_1 2} \ldots \sum_{k_{m-1}} W^{jk_{m-1}2^{m-1}} A(k_{m-1}k_{m-2}\ldots k_0).$$

Equation 6.2 consists of m nested summations. Since $W^N=1$,

$$W^{j2^{m-s}} = W^{j_{s-1}j_{s-2}\ldots j_0 \cdot 2^{m-s}},$$

so the innermost s summations depend only on the m binary variables j_0, ..., j_{s-1} and k_{m-s-1}, ..., k_0. Thus the innermost s summations represent a function from 0, ..., N–1 to the complex numbers; we represent this function as an array B_s of N complex numbers. B_s satisfies

(6.3) $B_0(k_{m-1}k_{m-2}\ldots k_0) = A(k_{m-1}k_{m-2}\ldots k_0)$,

$$B_s(j_0\ldots j_{s-1}k_{m-s-1}\ldots k_0) =$$

$$\sum_{k_{m-s}} B_{s-1}(j_0\ldots j_{s-2}k_{m-s}\ldots k_0) W^{j_{s-1}j_{s-2}\ldots j_0 \cdot 2^{m-s} \cdot k_{m-s}},$$

and

$$B_m(j_0\ldots j_{m-1}) = X(j_{m-1}\ldots j_0).$$

Equation 6.3 reveals how we can use the 4-pin shuffle to compute the DFT: We iteratively compute B_s for s = 1 to m. Iteration s results in B_s distributed on the output pins. To perform iteration s, we form the weighted sum of elements from B_{s-1} whose indices differ only in bit position number m–s. The 4-pin shuffle with N/2 processors provides exactly the data alignment we want, since shuffling an array s times causes the indices of the two numbers in each processor to differ only in bit position number m–s.

The following is a description of Pease's parallel FFT algorithm. The hardware is assumed to be a 4-pin shuffle with N/2 PEs. The machine operates in SIMD mode, and PEs differ only in that each processor $PE(p_{m-2}\ldots p_0)$ knows its own address $p_{m-2}\ldots p_0$.

Large-network Parallel FFT

Input: data items A(k)   k=0, ..., N–1
with A(k) on OPIN(k)

Output: the Fourier transform X(j) of A(k)
with $X(j_{m-1}\ldots j_0)$ on $OPIN(j_0\ldots j_{m-1})$
for s := 1 to m
begin
   SHUFFLE;
   $OPIN0 := IPIN0 + W^{0p_0\ldots p_{s-2}\cdot 2^{m-s}} \cdot IPIN1;$
   $OPIN1 := IPIN0 + W^{1p_0\ldots p_{s-2}\cdot 2^{m-s}} \cdot IPIN1;$
end

This algorithm can be proved correct by induction on the following loop invariant:

Immediately after shuffle number s,

and $\quad \begin{array}{l} B_{s-1}(j_0\ldots j_{s-2}0k_{m-s-1}\ldots k_0) \\ B_{s-1}(j_0\ldots j_{s-2}1k_{m-s-1}\ldots k_0) \end{array}$

are in processor $PE(k_{m-s-1}\ldots k_0 j_0\ldots j_{s-2})$ at pin positions

and $\quad \begin{array}{l} IPIN(k_{m-s-1}\ldots k_0 j_0\ldots j_{s-2}0) \\ IPIN(k_{m-s-1}\ldots k_0 j_0\ldots j_{s-2}1) \end{array}$

respectively. This processor then places

and $\quad \begin{array}{l} B_s(j_0\ldots j_{s-2}0k_{m-s-1}\ldots k_0) \\ B_s(j_0\ldots j_{s-2}1k_{m-s-1}\ldots k_0) \end{array}$

onto output pin positions

and $\quad \begin{array}{l} OPIN(k_{m-s-1}\ldots k_0 j_0\ldots j_{s-2}0) \\ OPIN(k_{m-s-1}\ldots k_0 j_0\ldots j_{s-2}1) \end{array}$

respectively.

6.3.2. Sorting on the Shuffle

Batcher's algorithm [11], as adapted by Stone [13], sorts N numbers in $\log^2 N$ passes through the N/2-processor 4-pin shuffle. After each shuffle, a processor either

1) Copies the two inputs directly to the two outputs.
2) Compares the two inputs and puts the lower on OPIN0 and the higher on OPIN1.

3) Compares the two inputs and puts the higher on OPIN0 and the lower on OPIN1.

6.3.3. Polynomial Evaluation on the Shuffle

Consider the problem of evaluating the (N–2)nd-degree polynomial

$$(6.4) \quad \sum_{i=0}^{N-2} a_i x^i$$

for given numbers x and a_i, i=0, ..., N–2. Horner's rule, which evaluates a polynomial by the scheme

$$(...((a_n x + a_{n-1})x + a_{n-2})x + ... + a_1)x + a_0,$$

is an optimal serial algorithm that requires exactly N–2 multiplications and N–2 additions. Stone [13] presents an algorithm for computing (6.4) with 2 log N passes through the N/2-processor 4-pin shuffle.

6.3.4. Finite-Difference Methods

The literature is full of proposals for the parallel execution of finite-difference calculations [6,28,44,46,47,48]. Often, the rectilinear problem grid is mapped one-to-one onto the rectilinear processor grid. At each time step, each processor communicates its values to and receives values from each of its nearest neighbors. This exchange provides each processor with the necessary values to compute the value of its point at the next time step.

6.4. Network Emulation

Definition. Suppose that $G = (V_G, E_G)$ and $H = (V_H, E_H)$ are graphs. We say that a function $f: V_H \longrightarrow V_G$ is an *emulation* of H by G if for every edge $(h_1, h_2) \in E_H$

$$f(h_1) = f(h_2) \text{ or } (f(h_1), f(h_2)) \in E_G.$$

Every emulation $f: V_H \longrightarrow V_G$ induces a mapping $f': E_H \longrightarrow V_G \sqcup E_G$ in a natural way:

$$f'(h_1, h_2) = (f(h_1), f(h_2)) \text{ if } (f(h_1), f(h_2)) \in E_G.$$

otherwise

$f'(h_1,h_2) = (f(h_1) = f(h_2)$.

We say that the node $g \in V_G$ *emulates* the nodes $f^{-1}(g)$, and that the edge $(g_1,g_2) \in E_G$ *emulates* the edges $f'^{-1}(g_1,g_2)$. If $|f^{-1}(g)|$ is the same for every $g \in V_G$, then we say that f is *computationally uniform*, and $|f^{-1}(g)|$ is the *computation factor* of f; if $|f^{-1}(e)|$ is the same for every $e \in E_G$, then we say that f is *exchange-uniform*, and $|f^{-1}(e)|$ is the *exchange factor* of f. If f is computationally uniform and exchange-uniform, and if the computation factor equals the exchange factor, then we say that f is *totally uniform*, and $|f^{-1}(g)|$ is the *emulation factor* of f.

If the graphs G and H are interconnection networks, then the existence of an emulation of H by G provides a way for the network G to emulate the actions of the network H. By analogy with the notion of quotient groups in abstract algebra, we call G a *quotient network*. The processor $g \in V_G$ is time-shared to emulate the group of processors $f^{-1}(g)$ in V_H, and the communications line $(g_1,g_2) \in E_G$ is time-multiplexed to emulate the communication lines $f'^{-1}(g_1,g_2)$ in E_H.

If the emulation of H by G is computationally uniform, then the processors in G can efficiently perform the actions of the processors of H: Since each processor in G emulates the same number of processors of H, all of the processors in G can proceed in unison and finish simultaneously. No processors sit idle while other overloaded processors finish their work. Likewise, if the emulation of H by G is exchange-uniform, then the communications lines in G can efficiently perform the actions of the communications lines of H: Since each communications line in G emulates the same number of communications lines of H, all of the data transfers in G can proceed in unison and finish simultaneously. No communications lines sit idle while other overloaded communications lines finish their work.

An emulation for each of the networks reviewed in section 2 is now presented. In each case, a large network H is emulated by a smaller network G of the same general interconnection scheme.

6.4.1. Perfect Shuffle

Suppose that H is a shuffle-exchange network with $N = 2^m$ processors. We will emulate this network with a 4-pin shuffle network of size $N/2$, and then emulate the 4-pin shuffle network with any 4-pin shuffle network of size a smaller power of two.

Theorem 6.1. The function $f(p_{m-1}...p_1p_0) = p_{m-1}...p_2p_1$ emulates the shuffle-exchange network of size N with the 4-pin shuffle of size $N/2$.

Proof. Suppose that $e = (h_1, h_2) \in E_H$. If e is an exchange connection, then $f(h_1) = f(h_2)$. If e is a shuffle connection, then

$$(f(h_1), f(h_2))$$
$$= (f(p_{m-1} \cdots p_1 p_0), f(p_{m-2} \cdots p_1 p_0 p_{m-1}))$$
$$= (p_{m-1} \cdots p_1, p_{m-2} \cdots p_1 p_0) \in E_G.$$

Q.E.D.

The emulation f is computationally uniform and exchange-uniform, but not totally uniform. The computation factor is two and the exchange factor is one.

Theorem 6.2. The function

$$f(p_{m+q-1} p_{m+q-2} \cdots p_q p_{q-1} \cdots p_0) = p_{m+q-1} p_{m+q-2} \cdots p_q$$

emulates the 4-pin shuffle of size $NP = 2^{m+q}$ with the 4-pin shuffle of size $N = 2^m$.

Proof. Let

$$(h_1, h_2)$$
$$= (p_{m+q-1} p_{m+q-2} \cdots p_q p_{q-1} \cdots p_0, \; p_{m+q-2} \cdots p_q p_{q-1} \cdots p_0 X)$$

be an edge in H. Then

$$(f(h_1), f(h_2))$$
$$= (p_{m+q-1} p_{m+q-2} \cdots p_q, \; p_{m+q-2} \cdots p_q p_{q-1}) \in E_G.$$

Q.E.D.

The emulation f is totally uniform, with emulation factor 2^q. Figure 6.1 illustrates a 4-pin shuffle with four PEs emulating a 4-pin shuffle with eight PEs.

6.4.2. Grid-Connected Network

The emulation of a large grid-connected network with a small one is fairly straightforward; we simply partition the large network into square regions.

Fig. 6.1. 4-PE 4-pin shuffle emulating 8-PE 4-pin shuffle.

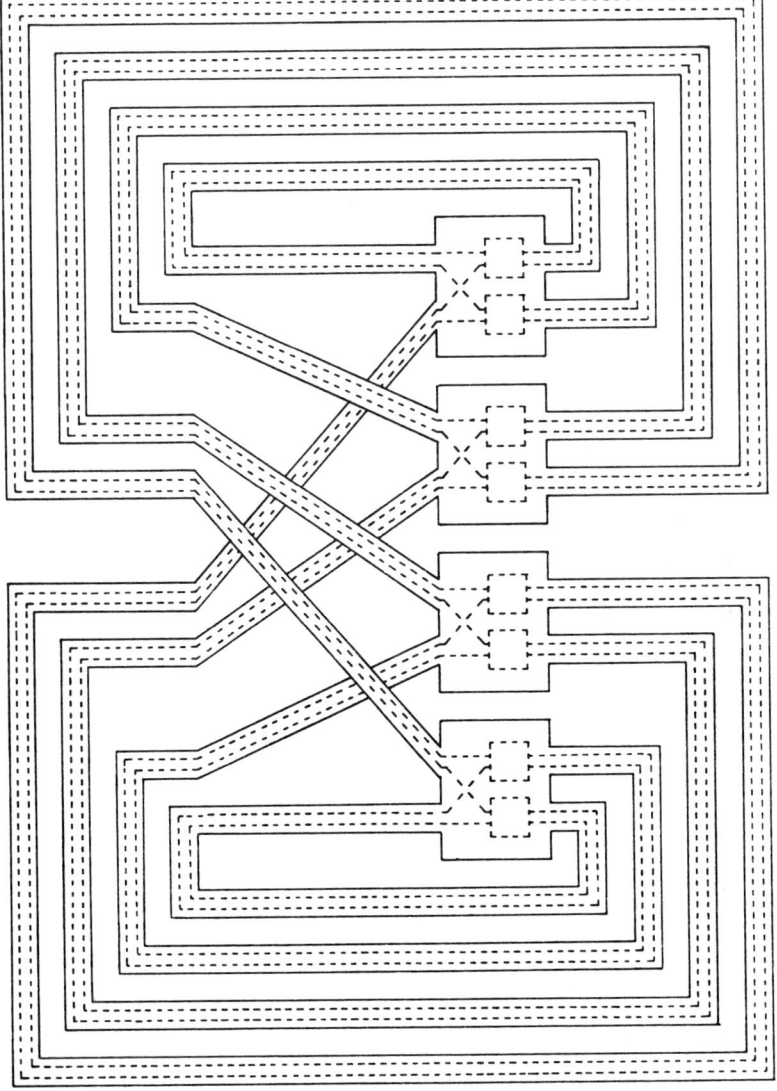

92   Quotient Networks

Theorem 6.3. The function

$$f(p_{r+s-1}\ldots p_r p_{r-1}\ldots p_0,\ q_{r+s-1}\ldots q_r q_{r-1}\ldots q_0)$$
$$= (p_{r+s-1}\ldots p_r,\ q_{r+s-1}\ldots q_r)$$

is an emulation of a grid-connected network of size 2^{2r+2s} by a grid-connected network of size 2^{2s}.

Proof. Suppose that $h = (h_1, h_2) = ((P_1, Q_1), (P_2, Q_2)) \in E_H$. We assume that h is a "North" connection, so that $P_1 = P_2$ and $Q_1 = Q_2 - 1$. A similar proof can be used when h is any of the other three grid connections. We can represent P_1, P_2, Q_1, and Q_2 as follows:

$$P_1 = P_2 = p_{r+s-1}\ldots p_r p_{r-1}\ldots p_0$$
$$Q_1 = Q_2 - 1 = q_{r+s-1}\ldots q_r q_{r-1}\ldots q_0$$

If incrementing Q_1 results in a carry into the top s bits in its binary representation, then

$$(f(h_1), f(h_2))$$
$$= (f(p_{r+s-1}\ldots p_r p_{r-1}\ldots p_0,\ q_{r+s-1}\ldots q_r q_{r-1}\ldots q_0),$$
$$\quad f(p_{r+s-1}\ldots p_r p_{r-1}\ldots p_0,\ q_{r+s-1}\ldots q_r q_{r-1}\ldots q_0 + 1))$$
$$= ((p_{r+s-1}\ldots p_r,\ q_{r+s-1}\ldots q_r),\ (p_{r+s-1}\ldots p_r,\ q_{r+s-1}\ldots q_r + 1))$$
$$\in E_G.$$

If not, then

$$f(h_1)$$
$$= f(P_1, Q_1)$$
$$= f(p_{r+s-1}\ldots p_r p_{r-1}\ldots p_0,\ q_{r+s-1}\ldots q_r q_{r-1}\ldots q_0)$$
$$= (p_{r+s-1}\ldots p_r,\ q_{r+s-1}\ldots q_r)$$
$$= f(p_{r+s-1}\ldots p_r p_{r-1}\ldots p_0,\ q_{r+s-1}\ldots q_r q_{r-1}\ldots q_0 + 1)$$
$$= f(P_2, Q_2)$$
$$= f(h_2).$$

Q.E.D.

The emulation f is computationally uniform and exchange-uniform, but not totally uniform. The computation factor is 2^{2r} and the exchange factor is 2^r. Figure 6.2 illustrates part of a grid-connected network emulating a grid-connected network that is four times larger.

In general, a k-dimensional grid may be emulated by a smaller k-dimensional grid. For a given r, the exchange factor is 2^r and the computation factor is 2^{kr}.

6.4.3. Cube

Theorem 6.4. The function

$$f(p_{m+q-1}p_{m+q-2}\cdots p_q p_{q-1}\cdots p_0) = p_{m+q-1}p_{m+q-2}\cdots p_q$$

emulates the cube of size $NP=2^{m+q}$ with the cube of size $N=2^m$.

Proof. Suppose that $(h_1, h_2) \in E_H$. Then h_1 and h_2 are of the form

$$h_1 = p_{m+q-1}\cdots p_i \cdots p_0$$

and

$$h_2 = p_{m+q-1}\cdots \overline{p}_i \cdots p_0$$

Fig. 6.2. Grid emulation with r=2.

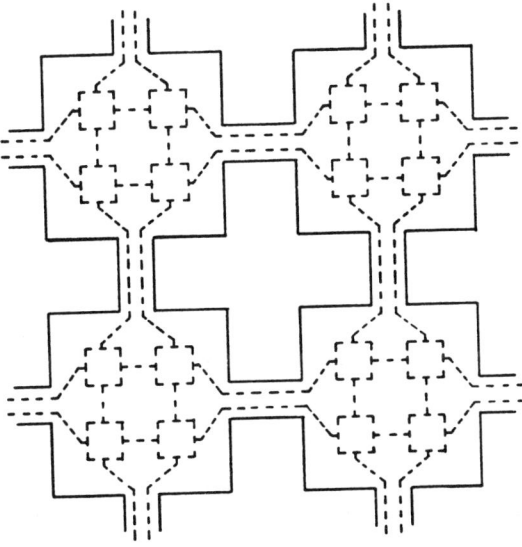

If $i \leq q$, then $f(h_1) = f(h_2)$. If $i > q$, then

$$f(h_1) = p_{m+q-1}\cdots p_i\cdots p_q$$

and

$$f(h_2) = p_{m+q-1}\cdots \overline{p}_i\cdots p_q$$

Hence $(f(h_1), f(h_2)) \in E_G$.

Q.E.D.

The emulation f is totally uniform, with emulation factor 2^q. Any function that discards q bits and permutes the remaining bits is also an emulation. Figure 6.3 illustrates a cube of size four emulating a cube of size eight.

6.4.4. PM2I

Theorem 6.5. The function

$$f(p_{m+q-1}p_{m+q-2}\cdots p_q p_{q-1}\cdots p_0) = p_{m+q-1}p_{m+q-2}\cdots p_q$$

emulates the PM2I network of size $NP=2^{m+q}$ with the PM2I network of size $N=2^m$.

Proof. Let $(h_1, h_2) \in E_H$. Hence h_1 and h_2 are of the form

$$h_1 = p_{m+q-1}p_{m+q-2}\cdots p_q p_{q-1}\cdots p_0$$

and

$$h_2 = p_{m+q-1}p_{m+q-2}\cdots p_q p_{q-1}\cdots p_0 + 2^i,$$

for some $0 \leq i < m$. If $i < q$ and if the addition of 2^i to h_1 does not cause a carry into the top m bits of its address, then $f(h_1) = f(h_2)$. Otherwise, if $i \geq q$ then

$$f(h_2) = f(h_1) + 2^{i-q},$$

and if $i < q$ then

Fig. 6.3. Two-dimensional cube emulating three-dimensional cube.

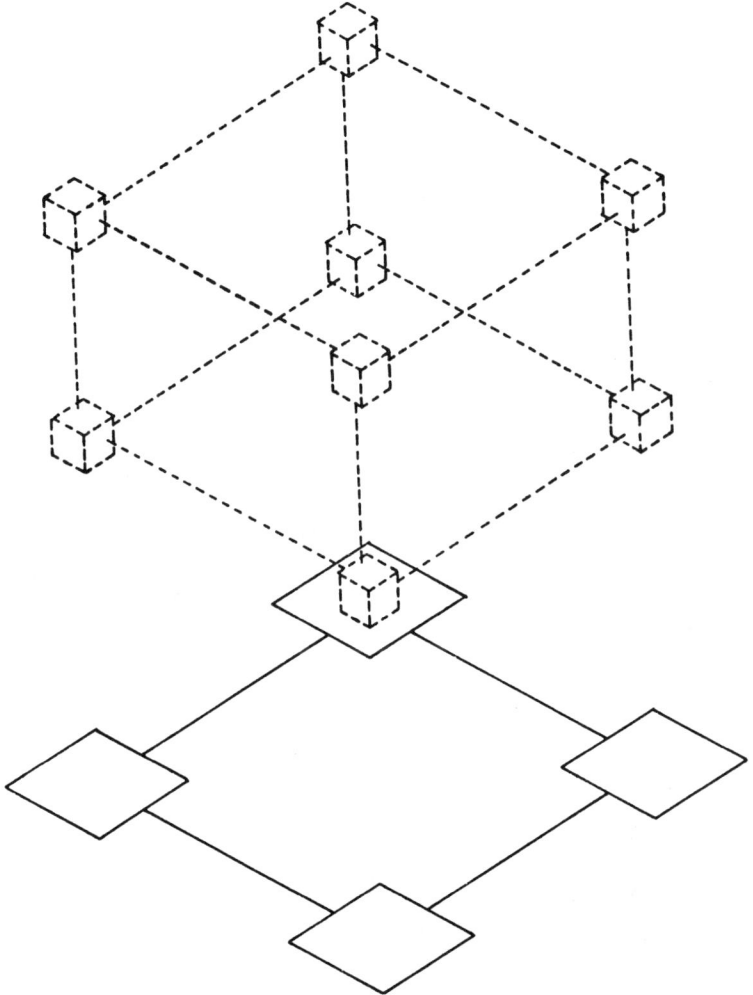

$f(h_2) = f(h_1) + 1$.

In either case, $(f(h_1), f(h_2)) \in E_G$.

Q.E.D.

The emulation f is computationally uniform, with computation factor 2^q. But f is not exchange-uniform, since each "+1" link in G emulates $2^{m+1}-1$ links in H, while every other link in G emulates 2^m links. Figure 6.4 illustrates a PM2I of size eight emulating a PM2I of size sixteen.

Fig. 6.4. PM2I of size eight emulating PM2I of size sixteen.

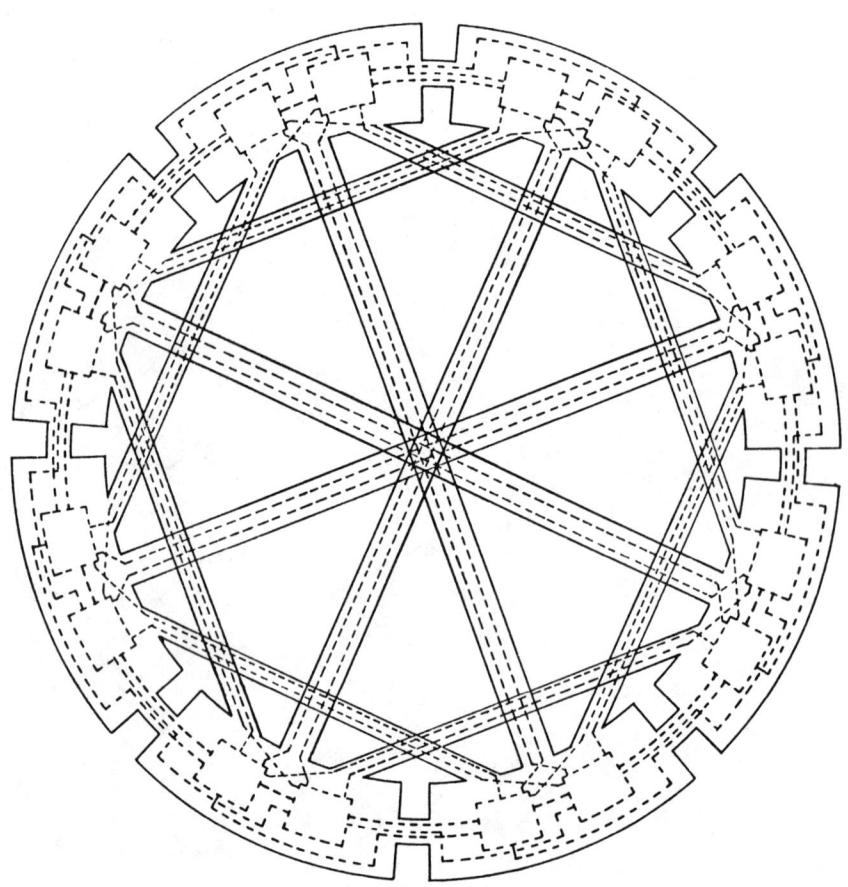

6.5. Some Resulting Algorithms

This section transforms the large-network algorithms of section 3 into quotient-network algorithms. Since the transformation is a fairly mechanical one, we present it in detail only for the FFT algorithm. For the other algorithms we only summarize the result.

6.5.1. Fast-Fourier Transform on the Shuffle

The FFT algorithm presented in section 3 consists of a loop executed m times. The body of the loop consists of a SHUFFLE followed by placing weighted sums of the input pins onto the output pins. We assume, as in section 3, that our network contains $N/2 = 2^{m-1}$ machines. We wish to compute the DFT of $N \cdot P = 2^{m+q}$ data items A(i) for $i = 0, \ldots, N \cdot P - 1$. We therefore emulate the actions of a 4 − p in network of size. Each processor represents the virtual pins of the processors it is emulating by arrays: The virtual $PE(k_{m+q-2}\ldots k_0)$ has pins OPIN0, OPIN1, IPIN0, and IPIN1. These pins are emulated on actual $PE(k_{m+q-2}\ldots k_q)$ at index $k_{q-1}\ldots k_0$ of arrays EOPIN0, EOPIN1, EIPIN0, and EIPIN1, respectively. Here is the quotient-network FFT algorithm:

Quotient-Network Parallel FFT

   Input: data items A(i), i=0, ..., N·P−1 such that

   $A(k_{m+q-1}\ldots k_{q+1}k_q\ldots k_0)$ is stored on machine
   $PE(k_{m+q-1}\ldots k_{q+1})$ in EOPIN$k_0[k_q\ldots k_1]$
   Output: The Discrete Fourier Transform X(i),
      i=0, ..., N·P−1 such that
   $X(k_0\ldots k_q k_{q+1}\ldots k_{m+q-1})$ is stored on machine
   $PE(k_{m+q-1}\ldots k_{q+1})$ in EOPIN$k_0[k_q\ldots k_1]$
   for s := 1 to m + q do
   begin
      { emulate SHUFFLE: }
      for j := 0 to 2^q −1 do
      begin { emulate $PE(P_{m-2}\ldots P_0 j_{q-1}\ldots j_0)$ }
         if j is even then
         begin
            OPIN0 := EOPIN0[j/2];
            OPIN1 := EOPIN0[j/2+2^{q-1}];
         end else

```
            begin
              OPIN0 := EOPIN1[⌊j/2⌋];
              OPIN1 := EOPIN0[⌊j/2⌋+2^(q-1)];
            end;
            SHUFFLE;
            EIPIN0[j] := IPIN0;
            EIPIN1[j] := IPIN1;
          end;
          { emulate computation: }
          for j := 0 to 2^q -1 do
          begin { emulate PE($P_{m-2}\ldots P_0 j_{q-1}\ldots j_0$) }
            $e_{m+q-2}\ldots e_0$ := $P_{m-2}\ldots P_0 j_{q-1}\ldots j_0$;
            EOPIN0[j] :=
              EIPIN0[j] + $W^{0e_0\ldots e_{s-2}\cdot 2^{m+q}}$ · EIPIN1[j];
            EOPIN1[j] :=
              EIPIN1[j] + $W^{1e_0\ldots e_{s-2}\cdot 2^{m+q}}$ · EIPIN1[j];
          end;
        end;
```

The original large-network FFT algorithm is optimal in the number of processors; that is, the speedup is proportional to the number of processors used. In the above example, the large-network FFT algorithm rests on top of an emulation, which rests on the actual hardware. Apart from the loop and indexing overhead needed to emulate the SHUFFLE step, the 2^{m-1}-processor network is 2^q times as slow as the 2^{m+q-1}-processor network. The loop and indexing overhead slows the algorithm down by only a constant factor, and could be eliminated entirely by unrolling the loops. Therefore the quotient-network algorithm is also optimal: We gain approximately N speedup with N processors. While the original large-network FFT algorithm performs log N operate-shuffle steps, the quotient-network algorithm performs 2^qlog N operate-shuffle steps.

6.5.2. Sorting on the Shuffle

As we mentioned above, Batcher's large-network algorithm sorts $N = 2^{m+q}$ numbers in $(m+q)^2$ operate-shuffle steps on a 4-pin shuffle with 2^{m+q-1} processors. A quotient-network version of this algorithm sorts the 2^{m+q} numbers in $2^q(m+q)^2$ operate-shuffle steps on a 4-pin shuffle with 2^{m-1} processors.

6.5.3. *Polynomial Evaluation on the Shuffle*

A 2^{m+q-1}-processor 4-pin shuffle network can evaluate a polynomial of degree $2^{m+q-1}-2$ in $2(m+q)$ operate-shuffle steps. The quotient-network version of this algorithm evaluates the same polynomial in $2^{q+1}(m+q)$ operate-shuffle steps with 2^{m-1} processors.

6.5.4. *Finite-Difference Methods*

A large-network algorithm that maps the 2^{2r+2s} points of a finite-difference grid one-to-one onto a 2^{2r+2s} grid-connected network must communicate each point at each time step to all four neighbors. The quotient-network version of this algorithm reduces the communication/computation ratio by communicating only the border points of a processor's region to that processor's neighbors. The *grid algorithm* of chapter 5 can now be seen as a quotient-network algorithm.

6.5.5. *Alpha-Beta Search*

One optimization of the tree-splitting algorithm mapped the top several layers of masters onto a single processor. For example, the root master and its two slaves can be processes on the root node of a processor tree. We can view this processor tree as a quotient network: The root node emulates the top three processors of a binary processor tree.

6.6. The Economics of Emulation

We can give several economic arguments in favor of solving large problems on small networks through emulation.

1) The cost of a word of storage is much smaller than the cost of a processor. This fact is independent of further increases in scale of integration. By adding extra storage at each processor in G, we increase the potential computation factor of an emulation; that is, we can emulate a larger H. We therefore increase the largest problem that the network can handle, with a much smaller increase in hardware cost than would be incurred by expanding G to H.
2) Suppose that a solution must meet a time constraint for a problem of size N. One processor cannot meet this constraint, but N processors (the network H) are much too expensive and much faster than

needed. An intermediate number of processors (the network G) emulating H may be fast enough and affordable.

3) Given a large-network algorithm, an emulation automatically produces a quotient-network algorithm to solve the same problem on a smaller machine. We achieve economy of thought by solving once and for all the "emulation problem": doing on a small machine what a large machine can do. Thereafter, we can deal exclusively with the simpler class of problems: data sets of size N on networks with N processors.

7
Conclusions and Future Directions

In literature, in art, in life, I think that the only conclusions worth coming to are one's own conclusions. If they march with the verdict of the connoisseurs, so much the better for connoisseurs; if they do not so march, so much the better for oneself.

A.C. Benson
From a College Window

This chapter reviews briefly the major contributions of this work. Where appropriate, areas that need further research are indicated.

7.1. Alpha-Beta Search

We have presented two parallel algorithms for implementing alpha-beta search on a tree of processors. The first, Palphabeta, divides work recursively among slave processors in a simple fashion. Under best-first ordering of the lookahead tree, Palphabeta achieves $k^{1/2}$ speedup with k processors. The second distributed algorithm, mwf, orders work to be done by slaves in a more sophisticated manner. Under best-first ordering of a chess lookahead tree, mwf achieves $k^{0.8}$ speedup with k processors. Our work with the parallel alpha-beta algorithms has led to the discovery of an optimization of the serial algorithm. This optimization, called Lalphabeta, is discussed in the appendix.

The question remains whether there exists an optimal parallel algorithm for alpha-beta search. We moved toward optimality in going from Palphabeta to mwf, but abandoned deep cutoffs along the way to simplify the analysis. Any optimal algorithm must achieve these cutoffs. Mwf should also be made more practical by increasing the number of processors that can be used to meet a time limit. Currently, mwf must run on a processor tree that is less than half the height of the lookahead tree being searched.

Both Palphabeta and mwf assumed that the logical topology of the multicomputer was a tree. This assumption allowed us to write down recursive relations that, when solved, gave the finishing time of the

102   Conclusions and Future Directions

algorithms. Unfortunately, this assumption increased slave idle time: Some of processor X's slaves are idle because X has more slaves than work, while at the same time processor Y has more work than slaves and could use those idle slaves. A distributed algorithm might view the collection of computers as a uniform pool. With this organization, idle time might be reduced. Further research should be directed toward the question of how to organize a pool of processors to perform parallel alpha-beta search.

7.2. Piecewise-Serial Iterative Methods

The Jacobi method is an iterative numerical technique for solving certain partial differential equations. We have shown how locally defined iterative methods give rise to natural multicomputer algorithms. In particular, the grid and the tree algorithms map parts of the problem onto individual processors. Each processor (or terminal processor in the case of a tree multicomputer) engages in serial computation on its region and communicates border values to its neighbors when those values become available.

Our analysis derives the running time of the grid and the tree algorithms with respect to per-message overhead, per-point communication time, and per-point computation time. As long as each machine has a significant amount of work to perform, message passing does not seriously degrade performance. The grid method is more efficient than the tree method, but the semi-synchronous optimization of the tree algorithm is more efficient than the grid algorithm if it is compute-bound. All three algorithms give nearly N-fold speedup with N machines on large problems; the speedup approaches the number of slave processors as the problem size goes to infinity. The efficiency of the tree algorithms depends on the tree fanout; we have shown that the optimal fanout is four. When a tree algorithm is used to solve problems in M dimensions, the optimal fanout is 2^M.

We have shown how to apply the tree algorithms to nonuniform regions both statically and dynamically. These modified algorithms shed load in a natural way from one slave to another when one has a larger area or more expensive computation.

The research reported here can be extended in several ways. We have assumed that the machines that compose the multicomputer are reliable. Experience shows that large assemblages of computers are very likely to have individual malfunctioning elements fairly often. If a machine should fail in such a way that it no longer receives or transmits data, the overall calculation should be able to continue. Consider the following grid of machines:

```
A  B  C  D
E  F  G  H
I  J  K  L
M  N  O  P
```

Let us suppose that J fails. In the case that each machine only deals with one mesh point, a reasonable adaptation is for machine K to use values from its neighbors, G, L, and O, as before, but to average values from F and N to estimate a value for J. Likewise, machine N will average values from I and K to substitute for missing values from J. However, if each machine is responsible for a sizable patch of mesh points, the recovery strategy is not so clear. Further research should investigate strategies for continuation when components fail. In the tree algorithm, dynamic region encroachment provides a natural algorithm for continuing when a single processor fails: All of a failed processor's load can be shifted to its siblings.

Dynamic region encroachment presents a method for redistributing subregions as the computation progresses to give each machine a similar amount of work. The question remains how often this redistribution should be done, since it requires a significant movement of data and an interruption of normal computation.

We expect that the advent of large-scale multicomputers will encourage further investigation into parallel algorithms for solving large numerical problems. For these algorithms to be efficient, they must be able to perform relatively large amounts of computation based on relatively small amounts of communication. As we saw in the case of the semi-synchronous tree algorithm, careful attention to the order of computation and communication can reduce communication cost.

7.3. Quotient Networks

By showing how to emulate a large interconnection network with a smaller network of the same topological family, we have presented a method for converting large-network algorithms into more practical small-network algorithms. It is important that the emulation does not in any way depend on the intended computation. Hence we can produce small-network versions of any large-network algorithm. The emulation itself produces no loss of efficiency, but allows us to perform the computation on a range of smaller machines.

7.4. Parallel Programming Proverbs

This section reviews some of the lessons learned while designing and implementing distributed algorithms.

7.4.1. Large Computation per Message

Distributed systems, unlike serial systems, must spend part of their time passing messages. Since message-passing in distributed systems can be time consuming, we are interested in algorithms that do as little communication as possible. Each message should invoke, or be a summary of, a large computation. Many of the algorithms presented in this study illustrate this principle.

1) Palphabeta and mwf use messages only to invoke or summarize the search of a large subtree. As the size of the lookahead tree increases, the amount of computation increases, but total message-passing time remains constant. Indeed, message-passing time does not even appear in the speedup formulas for our parallel alpha-beta algorithms, since these formulas represent limiting values as the lookahead tree becomes larger. If instead we had exploited parallelism in move generation or static evaluation, then total message-passing time would have remained proportional to computation time.
2) The grid and tree algorithms for locally defined iterative calculations were careful to partition the problem grid into segments with maximum area/boundary ratio. Since only boundary points need to be communicated to neighbors at each timestep, communication time is minimized. The area/boundary ratio increases linearly with the size of the problem grid. Hence the communication time does not appear in speedup figures for large problem grids.

7.4.2. Do Interesting Work First

The semi-synchronous algorithms for locally defined iterative calculations calculate new boundary points before new interior points. As soon as the new boundary points were available, they are sent to the master processor for distribution to other slaves. Only then does the calculation of interior points begin. By giving a head start to the dissemination of this "interesting information," the computation is speeded up, since the processor that is interested in the information is not kept waiting.

7.4.3. Do Mandatory Work First

The alpha-beta pruning technique prunes away some parts of the lookahead tree on the basis of information originating in other parts of the tree. Both of our parallel alpha-beta algorithms, Palphabeta and mwf, achieve parallelism by dividing the lookahead tree into subtrees and assigning each subtree to a separate task. Hence the results of some tasks can cause other tasks to be canceled. Suppose that task A cannot be canceled and that its result might cancel task B. Palphabeta is likely to lose the possible savings by executing A and B concurrently. Mwf achieves the savings, if possible, by executing A concurrently with some other noncancelable task. Mwf's smart behavior results in an improvement in speedup for searching best-first chess lookahead trees: With P processors mwf achieves $P^{0.8}$ speedup, as compared with $P^{0.5}$ speedup for Palphabeta.

7.4.4. Do Something

Into the life of every processor must come some idle time. If idleness is likely to happen often, we should arrange for something useful for the processor to do. For example, in the parallel alpha-beta algorithms, it can happen that

1) The queue of tasks is empty.
2) Some slaves are still busy with tasks.
3) Some slaves are idle.

We would like to give the idle slaves something useful to do, but the queue is empty. Worse, we cannot refill the queue until all the busy slaves finish. But with a little imagination we can find work: For example, we can assign the same subtrees that are still being worked on, but with smaller windows. Hence the same subtree might be searched concurrently by more than one processor, but with different-sized windows. If a processor with one of the smaller windows finishes first, then our strategy has paid off; we can send an alpha-beta update to the other processors working on the same tree. These processors then speedup or even terminate.

Appendix A

Some Optimizations of α-β Search

This appendix proposes three optimizations of the serial α-β algorithm.

A.1. Falphabeta

The first optimization, called Falphabeta for "fail-soft alpha-beta search," is completely riskless in the sense that it never searches more nodes than alphabeta. Although it requires a slight constant overhead, it results in a slight expected speedup whenever an initial window other than $(-\infty, +\infty)$ is used. Here is Falphabeta:

```
1   function Falphabeta(p: position; α,β: integer): integer;
2   begin integer m,i,t,d;
3       determine the successor positions p₁ ..., p_d;
4       if d = 0 then return(staticvalue(p));
5       m := - ∞;
6       for i := 1 to d do
7       begin
8           t := -Falphabeta(p_i,-β,-max(m,α));
9           if t > m then m := t;
10          if m ≥ β then return(m);
11      end;
12      return(m);
13  end;
```

Falphabeta differs from alphabeta only in that m has been initialized to $-\infty$ instead of α. In order to keep this change from affecting the third actual parameter to the recursive call to Falphabeta (line 8), "-m" is changed to "-max(m,α)." The computational overhead of repeatedly computing the maximum of m and α is the only added expense of Falphabeta. As mentioned in chapter 4, the value returned by the call to the original α-β procedure, alphabeta(p,α,β), obeys the following relation with respect to the true negamax value of a search tree:

If alphabeta $\leq \alpha$, then negamax(p) $\leq \alpha$,
If alphabeta $\geq \beta$, then negamax(p) $\geq \beta$,
if $\alpha <$ alphabeta $< \beta$ then negamax(p) = alphabeta.

Falphabeta obeys a stronger relation:

Appendix A

Theorem A. If p is the root node of a lookahead tree, and if α and β are integers satisfying $\alpha < \beta$, then the value Falphabeta returned by Falphabeta(p,α,β) satisfies:

If Falphabeta $\leq \alpha$, then negamax(p) $\leq$ Falphabeta,
If Falphabeta $\geq \beta$, then negamax(p) $\geq$ Falphabeta,
if $\alpha <$ Falphabeta $< \beta$ then negamax(p) = Falphabeta.

Proof. The relations clearly hold if p is a terminal node. Assume for the induction step that the relations hold for any tree of height k or less. Let p be the root of a tree of height k + 1. Let p_1, ..., p_d be the successors of p. Each p_i is the root of a tree of height k or less.

1) If

$$\text{Falphabeta}(p,\alpha,\beta) \leq \alpha,$$

then for all $1 \leq i \leq d$, we have

$$\text{Falphabeta}(p_i,-\beta,-\alpha) \geq -\alpha.$$

By the induction hypothesis, we have

$$\text{negamax}(p_i) \geq \text{Falphabeta}(p_i,-\beta,-\alpha).$$

Hence

$$\max_i -\text{negamax}(p_i) \leq \max_i -\text{Falphabeta}(p_i,-\beta,-\alpha).$$

Hence negamax(p) $\leq$ Falphabeta(p,α,β).

2) If Falphabeta$(p,\alpha,\beta) \geq \beta$, then there exists i such that

$$-\text{Falphabeta}(p_i,-\beta,-\alpha') = \text{Falphabeta}(p,\alpha,\beta) \geq \beta,$$

for some α' such that $\alpha \leq \alpha'$. By the induction hypothesis, we may conclude that

$$\text{negamax}(p_i) \leq \text{Falphabeta}(p_i,-\beta,-\alpha').$$

Hence negamax(p) = $\max_i$ -negamax$(p_i) \geq$ Falphabeta(p,α,β).

3) If $\alpha <$ Falphabeta$(p,\alpha,\beta) < \beta$, then let i be the smallest integer such that

$$-\text{Falphabeta}(p_i,-\beta,-\alpha') = \text{Falphabeta}(p,\alpha,\beta),$$

for some α' such that Falphabeta$(p,\alpha,\beta) > \alpha' \geq \alpha$. Hence

$$-\beta < \text{Falphabeta}(p_i,-\beta,-\alpha') < -\alpha'.$$

Therefore, by the induction hypothesis,

$$\text{negamax}(p_i) = \text{Falphabeta}(p_i,-\beta,-\alpha') = -\text{Falphabeta}(p,\alpha,\beta).$$

Since negamax(p) = negamax(p_i), we have

negamax(p) = Falphabeta(p,α,β);

Q.E.D.

Theorem A implies that Falphabeta can give a tighter bound than alphabeta on the true value of the tree when it fails high or low. Falphabeta "fails softer" than alphabeta. The extra information that Falphabeta gives can be used in two ways. First, this information is useful whenever the common wisdom "start with a tight window" is followed. If the tight window (α,β) causes the search to fail, the penalty of doing the entire search over again must be paid. With normal $\alpha-\beta$ search, this second search must be done with the window $(-\infty,\alpha)$ (if the original search failed low) or $(\beta, +\infty)$ (if the original search failed high). Falphabeta reduces this penalty: A low fail will sometimes return a number $k < \alpha$, and the second search can be started with the tighter window $(-\infty,k)$. We can expect a similar saving when a high fail occurs.

We need two definitions to explain the second use of Falphabeta. *Staged iteration* evaluates a lookahead tree to depth N by first searching to depths 2, 3, ..., N-1. After each stage, the principal continuation (the path the game would take if each player played optimally) is saved. The next stage begins its depth-first search by descending to the end of this path; whenever a node on the principal continuation is visited, its principal child is examined first. Staged iteration provides very reliable best-first move ordering at type-one nodes, so it actually decreases the number of nodes searched in chess programs.

Forward pruning, as opposed to $\alpha-\beta$ pruning, which is a form of backward pruning, cuts off a node of a tree before fully investigating any of its siblings. It is obvious that forward pruning can provide enormous savings in tree search. Unfortunately, forward pruning is very risky. No one has yet discovered how to perform forward pruning without occasionally pruning away the best move. (The very best chess programs do not perform forward pruning.) One of the reasons that forward pruning has not been successfully implemented is that when a poor move is evaluated after a better move, alphabeta assigns both the same score (except when the poor move is within two moves of the terminal node that produces the poor score). Falphabeta sometimes gives the poor move a more appropriate value, so it may provide a basis for reliably pruning the move during the next stage of a staged iteration.

A.2. Alphabeta

When alphabeta is recursively called on the last successor p_d, of the root of the entire tree, p, the current value $-\beta$ ($-\infty$) is passed as formal parameter α. Suppose that $-m-1$ is passed instead. If p_d is not the best move, then negamax(p_d) $\geq -m$, and alphabeta($p_d,-m-1,-m$) fails high as before. If p_d is the best move, then negamax(p_d) $\leq -m-1$, and so alphabeta($p_d,-m-1,-m$) fails low instead of succeeding. Nevertheless, the algorithm can still conclude that p_d is the best move, since its negamax value has been established to be lower than any other. The modified algorithm does not discover the value of the best move when that move is evaluated last. However, it still determines which move is best. This slight reduction in information can buy a time savings, since the evaluation of p_d has a very narrow window.

A parallel version of this technique was discussed in subsection 4.4.3. under the name "alpha-raising." The new algorithm will be called Lalphabeta, short for "last-move-with-minimal-window alpha-beta search."

```
1   function Lalphabeta (p: position; α,β: integer): integer;
2   begin integer m,i,t,d;
3       determine the successor positions p_i, ..., p_d;
```

```
4     if d = 0 then return(staticvalue(p));
5     m := α;
6     for i := 1 to d-1 do
7     begin
8         t := -alphabeta(p_i,-β,-m);
9         if t > m then m := t;
10        if m ≥ β then return(m);
11    end;
12    t := -alphabeta(p_d,-m-1,-m);
13    if t > m then m := t;
14    return(m);
15 end;
```

Lalphabeta provides an elegant solution to the forced-move problem: Programmers writing their first game-playing program often find to their amusement that alphabeta conducts a full-scale search even though only one move is available to the computer. Lalphabeta searches the one available move with the window $(\infty - 1, \infty)$. Besides greatly speeding up the search, Lalphabeta actually performs useful work in this case: It decides if it should resign!

A.3. Calphabeta

The third optimization, called Calphabeta because it is called only on nodes along the principal continuation, is a generalization of Lalphabeta, and profits from Falphabeta, but carries with it the risk that in certain cases more nodes will be examined.

```
1  function Calphabeta(p: position): integer;
2  begin integer m,i,t,d,;
3      generate the successors p_i, ..., p_d.
4      if d = 0 then return(staticvalue(p));
5      m = -Calphabeta(Pi);
6      for i := 2 to d do
       begin
8          t = -Falphabeta(p_i,-m-1,-m);
9          if t > m then m := -Falphabeta(p_i,-∞,-t)
10     end;
11     return(m);
12 end;
```

If Calphabeta evaluates the best move first at type one nodes, then all of the other subtrees are searched with a minimal window. On the other hand, every subtree that is better than its older siblings must be searched twice, resulting in more work. The first search, conducted with the minimal window, discovers that the subtree is the new best one, and really should not have been searched with the minimal window after all. The second search discovers the true value. It is important that the best move be evaluated first with high enough probability that the savings outweigh the penalties. Staged iteration can generate the best move first with high probability. If the principal line established for the (N-1)th stage is a prefix of the principal line for the Nth stage, then at the Nth stage virtually the entire tree is searched with a minimal window.

A.4. Measurements

To measure the improvement due to Lalphabeta and Calphabeta, four checkers games were played, during which the program made 46 moves. Each move selection was repeated six times, one for each of the six algorithms: alphabeta, Lalphabeta, Calphabeta, salphabeta, sLalphabeta, and sCalphabeta. Alphabeta, Lalphabeta, and Calphabeta have already been defined, and were done without staging. Salphabeta, sLalphabeta, and sCalphabeta are the staged versions of these three algorithms. During each of the 46*6 move selections, the number of nodes visited was counted, providing 46 values for alphabeta, Lalphabeta, Calphabeta, salphabeta, sLalphabeta, and sCalphabeta, and hence 46 values for the five derived quantities alphabeta/salphabeta, Lalphabeta/alphabeta, Calphabeta/alphabeta, sLalphabeta/salphabeta, and sCalphabeta/salphabeta.

Statistics for alphabeta/salphabeta are shown below. Checkers, unlike chess, does not profit from staging, possibly due to checker's small branching factor. On the average, alphabeta searched only 81% as many nodes as salphabeta.

Minimum	0.019
Maximum	2.768
Average	0.808
Standard Deviation	0.462

Statistics for Lalphabeta/alphabeta, Calphabeta/alphabeta, sLalphabeta/salphabeta, and sCalphabeta/salphabeta follow:

	Lalphabeta/alphabeta	Calphabeta/alphabeta
Minimum	0.881	0.666
Maximum	1.000	5.750
Average	0.987	1.163
Standard Deviation	0.024	0.868

	sLalphabeta/salphabeta	sCalphabeta/salphabeta
Minimum	0.899	0.696
Maximum	1.000	2.174
Average	0.988	0.960
Standard Deviation	0.023	0.227

As expected, staged iteration was crucial to making Calphabeta work at all; without staging, Calphabeta actually searched more nodes than alphabeta. However, the meaurements of sCalphabeta (Calphabeta with staging) are disappointing. SCalphabeta searched only 4% fewer nodes than salphabeta. However, experience with an algorithm similar to Calphabeta in the chess machine Belle has shown a speedup of about 1.5x [52]. The greater branching factor of chess may explain this discrepancy.

Lalphabeta and sLalphabeta, on the other hand, are unqualified (albeit small) successes. On the average, each searches about 1% fewer nodes than the corresponding standard algorithm. Although this improvement is not great, the optimization is clearly a good bargain, since its space overhead is insignificant and its time overhead is zero. Lalphabeta is never slower than alphabeta and sLalphabeta is never slower than salphabeta. Therefore, every game-playing program that uses $\alpha-\beta$ search should use some form of Lalphabeta.

Bibliography

[1] J. Backus, "Can programming be liberated from the von Neumann style? A functional style and its algebra of programs," *Communications of the ACM 21*, 8, pp. 613–41 (August 1978).
[2] W. J. Bouknight et al., "The Illiac IV system," *Proc. IEEE 60*, 4, pp. 369–88 (April 1972).
[3] W. A. Wulf and C. G. Bell, "C.mmp—a multiminiprocessor," *Proc. AFIPS 1972 Fall Joint Computer Conference 41, Part II*, pp. 765–77 (1972).
[4] J. A. Rudolph, "A production implementation of an associative array processor—Staran," *AFIPS Fall 1972 41* AFIPS Press, pp. 229–41 (1972).
[5] A. J. Evensen and J. L. Troy, "Introduction to the architecture of a 288 element PEPE," *Proc. 1973 Sagamore Conference on Parallel Processing*, (August 1973).
[6] P. M. Flanders, D. J. Hunt, S. F. Reddaway, and D. Parkinson, "Efficient high speed computing with the Distributed Array Processor," *Symposium on High Speed Computer and Algorithm Organization*, pp. 113–28 (1977).
[7] M. H. Solomon and R. A. Finkel, "The Roscoe distributed operating system," *Proc. 7th Symposium on Operating Systems Principles*, pp. 108–14 (December 1979).
[8] M. J. Flynn, "Very high-speed computing systems," *Proc. of the IEEE 54*, 12, pp. 1901–9 (December 1966).
[9] G. M. Baudet, *The Design and Analysis of Algorithms for Asynchronous Multiprocessors*, Department of Computer Science, Carnegie-Mellon University (April 1978).
[10] M. C. Pease, "An adaptation of the fast Fourier Transform for parallel processing," *Journal of the ACM 15*, 2, pp. 252–64 (April 1968).
[11] K. E. Batcher, "Sorting networks and their applications," *Proc. Spring Joint Comput. Conf. 32*, pp. 307–14 (1968).
[12] L. Csanky, "Fast parallel matrix inversion algorithm," *SIAM J. Computing 5*, 4, pp. 618–23 (December 1976).
[13] H. S. Stone, "Parallel processing with the perfect shuffle," *IEEE Transactions on Computers C-20*, 2, pp. 153–61 (February 1971).
[14] D. E. Muller and F. P. Preparata, "Bounds to complexities of networks for sorting and for switching," *Journal of the ACM 22*, 2, pp. 195–201 (April 1975).
[15] D. E. Knuth, *The Art of Computer Programming Volume 3—Sorting and Searching*, Addison-Wesley (1973).
[16] S. Even, "Parallelism in tape-sorting," *Communications of the ACM 17*, 4, pp. 202–4 (April 1974).
[17] L. G. Valiant, "Parallelism in Comparison Problems," *SIAM Journal of Computing 4*, 3, pp. 348–55 (September 1975).
[18] F. Gavril, "Merging with parallel processors," *Communications of the ACM 18*, 10, pp. 588–91 (October 1975).

Bibliography

[19] D. S. Hirschberg, "Fast Parallel Sorting Algorithms," *Communications of the ACM 21*, 8, pp. 657–61 (August 1978).

[20] F. P. Preparata, "New parallel-sorting schemes," *IEEE Transactions on Computers C-27*, 7, pp. 669–73 (July 1978).

[21] G. M. Baudet and D. Stevenson, "Optimal Sorting Algorithms for Parallel Computers," *IEEE Transactions on Computers C-27*, 1, pp. 84–87 (January 1978).

[22] C. D. Thompson and H. T. Kung, "Sorting on a mesh-connected parallel computer," *Communications of the ACM 20*, 4, pp. 263–71 (April 1977).

[23] H. S. Stone, "Sorting on STAR," *IEEE Transactions on Software Engineering SE-4*, 2, pp. 138–46 (March 1978).

[24] J. W. Cooley and J. W. Tukey, "An algorithm for the machine calculation of complex Fourier series," *Math. Comput. 19*, pp. 297–301 (April 1965).

[25] J. M. Lemme and J. R. Rice, "Speedup in parallel algorithms for adaptive quadrature," *Journal of the ACM 26*, 1, pp. 65–71 (January 1979).

[26] J. F. Traub, "Iterative solution of tridiagonal systems on parallel or vector computers," *Complexity of sequential and parallel numerical algorithms*, Academic Press (1973).

[27] H. S. Stone, "Parallel Tridiagonal Equation Solvers," *ACM Transactions on Mathematical Software 1*, 4, pp. 289–307 (December 1975).

[28] H. S. Stone, "Problems of Parallel Computation," *Complexity of Sequential and Parallel Numerical Algorithms*, Academic Press (1973).

[29] S. C. Chen, D. J. Kuck, and A. H. Sameh, "Practical Parallel Band Triangular System Solvers," *ACM Transactions on Mathematical Software 4*, 3, pp. 270–77 (September 1978).

[30] W. M. Gentleman, "Some complexity results for matrix computations on parallel processors," *Journal of the ACM 25*, 1, pp. 112–15 (January 1978).

[31] F. P. Preparata and D. V. Sarwate, "An improved parallel processor bound in fast matrix inversion," *Information Processing Letters 7*, 3, pp. 148–50 (April 1978).

[32] R. A. Finkel, M. H. Solomon, and M. L. Horowitz, "Distributed algorithms for global structuring," *Proc. National Computer Conference 48*, AFIPS Press, pp. 455–60 (June 1979).

[33] E. Chang, "An introduction to echo algorithms," *Proc. 1st International Conference on Distributed Computers*, pp. 193–98 (October 1979).

[34] D. S. Hirschberg, A. K. Chandra, and D. V. Sarwate, "Computing connected components on parallel computers," *Communications of the ACM 22*, 8, pp. 461–64 (August 1979).

[35] C. D. Savage, "Parallel algorithms for graph theoretic problems," Computer Science Department, U. of Illinois, Urbana, Ill. (August 1977) Ph.D. Thesis.

[36] H. J. Berliner, "A chronology of computer chess and its literature," *Artificial Intelligence 10*, pp. 201–14 (April 1978).

[37] S. G. Akl, D. T. Barnard, and R. J. Doran, "Simulation and analysis in deriving time and storage requirements for a parallel alpha-beta algorithm," *Proc. 1980 International Conference on Parallel Processing*, pp. 231–34 (August 1980).

[38] D. E. Knuth and R. W. Moore, "An analysis of alpha-beta pruning," *Artificial Intelligence 6*, 4, pp. 293–326 (Winter 1975).

[39] A. L. Samuel, "Some studies in machine learning using the game of checkers, II—recent progress," *IBM Journal of Research and Development*, p. 601–17 (November 1967).

[40] R. Nevanlinna and V. Paatero, *Introduction to Complex Analysis*, Addison-Wesley (1969).

[41] Oskar Perron, "Zur Theorie der Matrices," *Math. Ann. 64*, pp. 248–63 (1907).

[42] P. Brinch Hansen, *Operating System Principles*, Prentice-Hall (1973).

[43] D. M. Young, *Iterative Solution of Large Linear Systems*, Academic Press (1971).
[44] J. L. Rosenfeld, "A case study in programming for parallel-processors," *CACM 12*, 12, pp. 645-55 (December 1969).
[45] C. F. R. Weiman and C. E. Grosch, "Parallel processing research in computer science: Relevance to the design of a Navier-Stokes computer," *Proc. 1977 International Conference on Parallel Processing*, pp. 175-82 (August 1977).
[46] H. O. Welch, "Numerical weather prediction in the Pepe parallel processor," *Proc. 1977 International Conference on Parallel Processing*, pp. 186-92 (August 1977).
[47] D. Chazan and W. Mirankar, "Chaotic relaxation," *Linear Algebra and Appl. 2*, pp. 199-222 (1969).
[48] G. M. Baudet, "Asynchronous iterative methods for multiprocessors," *Journal of the ACM 25*, 2, pp. 226-44 (April 1978).
[49] J. L. Bentley and H. T. Kung, "A tree machine for searching problems," *Proc. 1979 International Conference on Parallel Processing*, pp. 257-66 (August 1979).
[50] L. J. Siegel, P. T. Mueller, and H. J. Siegel, "FFT algorithms for SIMD machines," *Proc. Allerton Conference on Communication, Control, and Computing*, pp. 1006-15 (October 1979).
[51] H. J. Siegel, "Analysis techniques for SIMD machine interconnection networks and the effects of processor address masks," *IEEE Transactions on Computers C-26*, 2, pp. 153-61 (February 1977).
[52] K. Thompson, personal communication, (November 1981).

Index

Accuracy property, 14; failing high, 14; failing low, 14; success, 14
Akl, S.G., 11, 15, 41
Alpha-beta algorithm, 11-15; optimizations, 109-11
Alpha-beta window, 14
Anderson, K., ix
Arachne (distributed operating system), ix, 1, 19-20, 74
Auden, W.H., 83

Backus, J., 1
Barnard, D.T., 11, 15, 41
Batcher, K.E., 5, 7, 87
Baudet, G.M., ix, 6, 15, 58, 83
Belle (chess machine), 111
Benson, A.C., 101
Best-first ordering, 24

C.mmp, 1, 4, 58-59
CDC Star computer, 7
Change, E., 8
Chaotic relaxation, 58-59
Chazan, D., 58
Checkers, 19
Chen, S.C., 8
Cray 2, 55
Cray, S., 55
Csanky, L., 8

Dewey decimal system, 24
Dirichlet problem, 56; discrete version, 56
Discrete Fourier Transform (DFT), 7, 85
Distributed Array Processor (DAP), 58
Doran, R.J., 11, 15, 41

Economics of Emulation, 99-100
Economy of thought, 83, 100
Even, S., 6

Fail-soft alpha-beta search (Falphabeta), 107-9
Fernbach, S., 55

Finkel, R.A., ix, 8
Flanders, P.M., 7, 58
Flynn, M.J., 1

Gauss-Seidel method, 57, 72
Gavril, F., 6
Gentleman, W.M., 8
Geophysical Fluid Dynamics Laboratory benchmark, 58
Global structuring, 8-9
Goodman, J., ix
Grid algorithm, 60; efficiency, 71

Hirschberg, D.S., 6, 9

IBM 360/195 computer, 58
Illiac IV, 1, 84
Interconnection networks, 84-85; cube, 85; four-pin shuffle, 84, 86; grid-connected, 84-85; plus-minus network, 84-85; shuffle-exchange, 84
Interesting work first, 104-5

Jacobi method, 56-57
John the Baptist, 11

Knuth, D.E., 6, 13, 30, 31

LSI-11 computer, 19
Large computation per message, 104
Last-move-with-minimal-window alpha-beta search (Lalphabeta), 109-10
Lawless, S.A., ix
Leland, M., ix
Leland, W.E., ix
Lemme, J.M., 7
Lookahead tree, 12; branching factor, 14; degree, 12; level of a node, 12; position, 12; predecessor, 12; root node, 12; successor, 12
Lower-triangular systems of equations, 8

Index

Macaulay, Lord, 11
Mandatory-work-first, 15-16, 105
Matrix inversion, 7-8
Matrix multiplication, 7-8
Measurement of communication time: per-point time, 72; per-message time, 74
Minimax algorithm, 12
Mirankar, W., 58
Moore, R.W., 42
Mueller, P.T., 83
Muller, D.E., 5
Multicomputer, 5-6; master, 15; maximal processor tree, 53-54; scheduling tree machine, 74-75; slave, 15; tree, 14; versus more powerful serial computer, 51-53
Multiple Instruction Stream, Multiple Data Stream (MIMD), 1-2
Multiprocessor, 1
Mwf algorithm, 41-51; best-first ordering, 44-47; comparison with Palphabeta, 50-51; other orderings, 49-50; worst-first ordering, 47-49

Navier-Stokes equation, 58
Negamax algorithm, 13
Network emulation, 88-96; computationally uniform, 89; computation factor, 89

Odd-even reduction algorithm, 8

Pairing algorithms, 8
Palphabeta algorithm, 23-29; analysis of speedup, 23-29; comparison to tree-splitting algorithm, 29; best-first ordering, 24-28; worst-first ordering, 23-24
Parallel Algorithms, 3-9; adaptive quadrature, 7; alpha-beta search, 11-12; comparison-exchange networks, 5-6; efficiency, 4; fast-Fourier transform (FFT), 7, 85; finite-difference methods, 88; graph-theoretic, 9; matrix methods, 7-8; multiprocessor sorting methods, 5; numerical methods, 7; polynomial evaluation, 88; practicality, 4, SIMD sorting methods, 5; sorting and merging, 5-7; speedup, 3; tape sorting, 6; vector sorting, 7; Crosspoint switch, 4
Parallel aspiration search, 15
Parallel processors, 6
Pascal programming language, 13
Pbound algorithm, 30; analysis of speedup under random order, 29-41
Pease, M.C., 7, 85
Pepe parallel processor, 58
Ph.D. thesis, average, 5
Piecewise-serial iterative methods, 55-82
Preparata, F.P., 6, 8

Principal-continuation alpha-beta search (Calphabeta), 110-11

Quicksort, 7
Quotient networks, 83-100; definition, 88; cube, 93-94; emulation factor, 89; exchange factor, 89; exchange-uniform, 89; four-pin shuffle, 90; grid-connected, 90-92, 93, plus-minus network, 96; shuffle-exchange, 89
Quotient-network algorithms, 97-99; alpha-beta search, 99; FFT, 97-98; finite-difference methods, 99; polynomial evaluation, 90; sorting, 98-99

Rosenfeld, J.L., 58

Savage, C.D., 9
Semi-synchronous tree algorithm, 68-71; compute-bound, 70; efficiency, 71; exchange-bound, 70
Siegel, H.J., 83
Siegel, L.J., 83
Single Instruction Stream, Multiple Data Stream (SIMD), 1-2
Smith, S.D., ix
Solomon, M., ix
Space Shuttle, 1
Spanning tree algorithms, 8
Static value, 12
Stevenson, D., 83
Stone, H.S., 5, 7, 8, 58, 87, 88
Strikwerda, J., ix
Successive over-relaxation (SOR), 57
Synchronous tree algorithm, 61-70; dynamic region encroachment, 75; efficiency, 71-72; optimal fanout, 63-66; M-dimension problem space, 67-68; nonoverlap of send and receive, 66-67; skewed quadrants, 75; static nonuniform regions, 75

Thompson, C.D., 6
Tips for processor-tree architects, 51-54
Traub, J.F., 8
Tree-splitting algorithm, 16-22; alpha-raising, 19; asynchronous updates, 16-17, 18-19; interior algorithm, 17-19; leaf algorithm, 16-17; measurements, 19-21; optimizations, 21-22

Uhr, L., ix

Von Neumann bottleneck, 1, 4

Weiman, C.F.R., 58
Welch, H.O., 58
Whitehead, A.N., 83
Worst-first ordering, 23-24

RAYMOND H. FOGLER LIBRARY
DATE DUE

BOOKS